Coloring Squared

Third Grade

A pixel art coloring book to teach and review 3rd grade math concepts.

By Cameron Krantzman

Table of Contents

Math Concepts

Math Facts

Name: Date:

Basic Multiplication

City

3x8	3x9	5x6	8x3	6x4	3x9	3x7	7x3	4x7	7x4	10x3	3x10	9x3	3x7	6x5	5x4	9x4	7x7	4x8
3x10	9x3	5x6	5x6	5x5	3x7	6x5	10x3	9x3	3x8	5x5	10x3	8x3	5x6	4x4	3x4	9x2	9x2	5x3
3x7	3x10	6x4	8x3	3x7	5x6	4x6	10x3	7x3	3x7	3x8	7x4	5x5	4x7	6x3	5x8	4x8	6x8	8x5
10x3	3x9	3x10	8x3	9x3	10x3	3x10	9x3	3x10	5x5	6x4	8x3	10x3	4x7	6x3	8x6	3x10	5x10	3x8
3x9	6x5	5x6	3x4	8x2	4x4	6x3	2x10	2x10	2x8	6x5	5x5	10x3	3x9	6x2	8x6	5x8	8x4	10x4
6x5	7x3	3x6	3x2	9x1	2x1	8x1	2x1	2x8	5x3	6x5	3x8	6x4	5x5	2x6	8x4	3x9	4x10	7x3
3x9	3x6	1x10	1x8	1x4	2x1	5x2	2x9	8x0	2x7	3x8	9x3	7x3	3x8	3x5	6x6	5x9	4x9	8x6
3x5	9x2	9x2	3x5	2x10	5x4	4x3	7x0	0x0	8x2	6x5	5x5	9x3	8x3	4x5	7x6	9x3	5x10	6x4
3x4	1x10	3x1	7x1	2x4	1x1	9x2	10x0	1x0	10x2	9x3	5x6	6x5	3x10	10x2	4x9	10x4	7x5	4x10
2x10	9x1	7x4	2x3	5x6	1x4	2x8	2x0	0x4	4x5	6x4	4x6	7x3	9x3	2x10	9x5	3x8	5x9	3x9
2x8	2x1	3x2	1x1	4x2	1x5	2x10	1x0	0x1	4x3	7x2	8x2	5x3	3x5	8x2	7x5	7x7	8x6	5x10
8x2	5x1	8x3	4x2	7x3	9x1	6x3	0x0	10x2	0x1	0x4	0x5	10x0	0x6	5x3	7x7	8x3	6x6	9x3
2x8	1x3	1x9	1x9	1x7	4x1	8x2	6x2	1x0	0x1	0x8	0x7	1x0	1x0	8x2	8x4	5x8	8x4	6x6
2x9	4x1	5x6	1x7	6x4	7x1	4x5	4x5	3x6	4x5	4x4	5x3	3x5	3x4	2x8	4x10	8x3	4x10	3x7
8x2	1x4	2x4	2x1	9x1	1x10	4x4	1x3	1x6	1x6	1x3	3x2	8x1	5x1	6x2	10x5	9x4	10x5	5x10
10x2	4x2	3x9	1x4	6x4	1x1	3x6	1x5	7x3	9x1	4x6	9x1	7x4	1x6	8x2	9x5	4x6	10x4	3x10
2x8	3x1	10x1	8x1	1x1	2x3	3x5	1x10	5x2	5x2	2x5	2x4	4x2	4x1	2x10	5x10	10x5	4x9	7x5
2x6	1x3	9x0	9x0	2x3	2x3	6x3	5x1	10x1	1x2	2x5	0x3	0x3	1x6	2x8	6x8	6x0	0x6	8x5
8x2	10x2	4x5	8x2	5x4	9x2	7x2	2x7	5x4	7x2	2x9	5x4	4x3	5x3	2x10	5x3	10x2	2x10	5x3
2x0	0x9	5x0	0x6	0x6	0x8	0x0	0x6	0x6	0x10	4x0	0x4	10x0	0x9	0x2	0x9	0x2	2x0	8x0

Key:

0	Dark Gray
1-10	Light Gray
11-20	Black
21-30	Light Blue
31-50	Dark Blue

Name: Date:

Basic Multiplication

Cthulu

10x4	6x7	8x4	4x8	6x8	10x2	4x4	3x6	6x2	4x4	2x7	2x9	5x8	4x8	9x5	5x9	10x5	4x10	7x6
10x4	9x5	6x8	8x4	2x10	8x1	4x1	4x2	6x1	9x1	1x2	6x3	7x6	9x5	4x9	8x5	4x9	8x5	4x10
10x5	4x8	7x6	8x4	2x6	8x1	10x3	9x3	4x6	3x10	7x3	3x2	2x6	8x5	4x10	4x9	8x5	10x4	8x6
9x5	7x6	8x5	4x9	5x3	8x1	9x3	6x4	4x7	7x4	9x3	5x2	4x5	9x4	7x7	10x4	5x8	7x5	6x8
9x4	5x7	8x6	7x5	9x2	9x1			9x3			5x1	6x3	7x6	10x5	8x4	5x7	10x5	5x8
6x6	4x8	6x6	7x7	4x3	7x1		2x3	7x3	2x5		3x1	9x2	7x7	4x8	9x5	5x7	9x5	4x9
5x7	7x7	5x9	4x10	9x2	2x3	4x7	10x3	9x3	6x4	7x4	1x6	4x5	5x10	5x7	9x4	4x10	7x6	6x7
4x9	8x6	8x5	4x8	2x10	9x1	5x2	7x3	3x7	7x4	1x1	1x4	4x3	8x4	6x7	10x5	9x5	10x5	4x9
7x5	9x4	5x8	8x4	2x9	7x1	7x1	9x3	9x3	5x6	1x2	1x2	1x7	10x2	10x5	8x5	6x7	6x8	7x7
5x7	5x9	6x6	8x6	8x2	7x1	7x4	7x3	4x7	5x6	10x3	4x1	2x5	10x1	2x10	3x8	7x3	6x7	7x5
7x7	4x8	8x5	2x6	4x4	7x3	3x9	3x7	5x6	6x4	7x3	3x10	3x4	2x10	3x1	5x5	6x4	5x1	5x9
9x5	8x6	5x8	6x3	2x9	3x9	10x2	8x2	3x10	5x4	6x4	6x3	5x6	3x4	7x1	9x2	5x7	5x2	6x6
6x8	9x5	7x5	2x10	2x7	5x5	8x2	6x4	1x3	2x7	6x4	5x4	5x5	2x9	4x2	8x1	5x1	2x5	10x5
10x4	10x4	3x4	4x5	3x4	8x3	10x2	4x6	6x1	3x7	4x4	10x2	3x6	5x4	3x2	2x6	7x2	10x1	8x6
5x10	7x7	5x3	9x2	4x4	9x2	6x2	2x6	1x2	2x9	2x6	2x7	6x3	4x9	6x2	2x0	3x0	2x6	4x8
9x5	6x8	6x5	3x9	10x4	10x2	3x6	2x6	9x1	4x4	2x8	7x2	4x3	5x8	3x4	0x2	4x0	10x2	5x9
8x5	9x4	4x7	7x4	6x6	2x6	9x2	4x4	3x6	2x4	2x7	6x2	10x2	6x8	4x8	4x4	4x3	8x6	8x5
10x4	4x9	5x7	8x6	8x6	2x7	4x5	4x3	6x3	1x2	5x4	6x3	7x7	8x6	9x4	6x6	9x5	9x4	9x4
7x6	7x7	10x4	5x7	8x5	3x6	3x5	6x2	10x2	2x1	9x2	9x2	6x8	4x8	8x5	5x10	4x9	7x5	9x4
4x10	10x5	6x6	8x4	4x9	5x4	6x2	6x2	10x2	5x1	2x9	10x2	4x9	10x5	4x8	6x7	9x4	5x10	5x8

Key:

0	Yellow
1-10	Black
11-20	Brown
21-30	Pink
31-50	Gray

*Blank squares are white

Name: Date:

Basic Multiplication

Apple Tree

3x9	4x7	3x8	8x3	4x6	7x4	3x7	8x3	4x7	4x7	6x5	1x3	8x5	8x5	4x9	10x4	5x7	8x6	5x9
5x6	3x10	7x3	9x3	6x5	9x3	7x4	7x3	6x4	5x6	7x4	4x2	9x4	6x7	5x7	7x7	9x5	8x4	6x6
5x5	10x3	6x4	8x3	5x6	4x6	7x3	8x3	5x6	1x4	1x1	10x5	4x10	10x5	7x6	8x6	8x6	5x7	8x4
1x2	7x4	7x4	3x7	8x3	10x3	5x5	4x7	1x4	9x5	4x8	5x10	6x8	6x7	6x8	6x7	4x9	8x4	5x10
4x9	2x1	6x1	9x3	7x4	3x10	1x1	1x4	7x5	8x4	1x6	8x5	4x8	2x5	1x6	2x1	2x4	6x8	5x7
7x7	2x3	10x0	1x9	1x7	1x1	6x0	4x1	8x4	10x1	0x7	1x7	2x1	10x3	3x7	6x5	3x10	1x7	7x7
5x10	4x2	0x6	0x10	3x0	0x7	8x0	1x9	4x10	4x2	0x9	0x3	9x1	3x7	6x4	3x10	4x6	3x9	1x1
8x5	1x4	6x0	6x0	0x10	0x0	9x1	6x1	4x2	5x10	5x1	7x0	1x5	3x8	2x5	3x1	1x1	1x8	10x4
8x4	3x2	0x10	2x0	4x0	10x1	2x9	8x2	2x7	4x2	1x7	0x5	1x5	1x9	2x7	3x4	5x4	10x2	2x3
10x5	10x1	2x0	9x0	1x5	7x2		5x3	9x2	6x2	1x10	0x7	5x1	6x3	5x3	2x9			3x1
10x5	3x1	0x2	6x1	9x2		4x5	2x6	2x7	2x6	2x6	1x4	4x3	3x5	7x2	2x8		1x10	10x1
10x5	1x9	0x4	4x1	2x6	4x5	2x7	5x3	2x7	2x10	2x8	3x6	9x2	3x6	2x6			1x5	8x6
4x8	1x5	0x1	7x1	5x3	2x9	2x6	3x5	5x4	6x2	4x3	5x4	4x5	3x4	4x5			5x1	6x8
9x5	5x2	6x0	4x2	8x2	2x8	5x4	3x5	3x6	3x4	5x3	9x2	3x5	2x8	4x3	2x8			2x2
10x5	4x2	4x0	0x6	1x6	8x2	7x2	3x4	9x2	9x2	9x2	9x2	3x6	6x2	2x7	4x5		3x2	8x6
2x5	0x9	0x2	0x8	1x3	6x3	7x2	4x3	7x2	4x3	2x9	2x8	2x10	3x5	3x5	5x3	2x8	6x1	4x8
10x0	0x0	7x0	0x1	0x6	1x2	4x4	2x7	2x7	6x2	6x2	5x4	4x5	6x3	3x5	2x6	7x1	7x6	4x9
0x7	4x1	2x0	8x0	4x2	2x2	5x3	8x2	6x2	6x2	6x2	2x6	4x5	2x9	4x3	1x7	9x3	5x5	7x4
3x2	0x10	1x3	10x1	3x10	10x3	1x5	1x7	9x2	8x2	4x1	1x10	3x4	3x6	1x8	3x10	3x10	4x7	7x4
3x7	5x2	8x3	3x8	6x4	5x5	4x6	3x8	1x2	1x1	3x9	3x8	2x4	8x1	6x4	7x4	3x8	3x9	4x6

Key:

0	Brown
1-10	Black
11-20	Red
21-30	Green
31-50	Blue

*Blank squares are white

Name: Date:

Basic Multiplication

Scooter

8x9	7x9	9x8	8x8	8x10	8x8	9x9	9x10	7x9	8x10	8x8	9x10	10x8	10x10	8x9	10x7	9x10	10x8	8x10
8x10	10x8	9x7	7x9	10x8	8x9	10x9	10x10	8x9	10x9	9x10	9x9	8x9	7x9	8x9	8x8	8x8	9x9	10x10
10x9	10x10	10x10	9x8	7x9	8x8	10x7	2x1	9x1	1x2	1x9	9x8	9x7	8x10	8x10	8x9	10x7	8x10	8x9
8x9	10x9	9x8	9x9	8x8	10x8	2x1	3x1	6x3	6x1	4x5	8x10	9x9	9x9	7x10	9x9	8x10	10x7	8x8
10x10	8x9	7x10	8x10	8x9	7x9	9x10	0x4	10x0	1x0	0x2	9x10	10x10	7x9	7x10	8x10	10x9	10x8	10x9
9x7	9x8	8x8	7x9	10x10	10x9	7x10	0x4	4x0	0x4	8x0	9x9	9x8	8x10	10x10	10x10	8x10	7x10	10x8
8x8	10x7	8x10	5x9	0x5	0x7	6x2	4x4	0x4	6x0	0x7	10x8	9x8	9x10	9x9	9x7	10x10	9x9	9x8
8x9	9x9	10x10	8x6	7x6	4x0	9x0	8x0	6x3	5x3	2x10	9x7	8x10	9x9	9x8	10x9	8x8	8x9	9x9
7x10	7x10	9x9	6x10	10x8	9x10	8x9	10x10	6x2	5x3	2x3	9x8	9x9	8x10	10x9	9x7	8x8	10x10	8x10
7x10	7x9	9x8	6x7	9x9	9x7	9x7	7x9	5x2	2x8	2x7	9x8	8x9	10x9	9x9	10x8	10x9	9x8	7x10
10x10	8x10	7x9	7x6	8x9	10x10	9x10	9x9	10x9			8x10	10x9	8x10	8x9	7x9	10x8	10x8	10x10
8x8	8x10	10x10	10x6	9x8	8x9	9x7	7x9	10x7			10x9	10x10	10x7	8x8	7x10	9x9	9x10	8x10
8x9	8x10	9x9	7x6	8x8	10x7	8x10	8x9	9x8			10x8	7x9	9x7	7x10	10x9	8x9	9x8	7x9
10x10	10x8	10x7	5x9	7x9	10x10	8x8	7x10	7x9			8x9	7x10	7x9	9x8	9x8	10x8	10x10	8x8
10x7	10x8	10x9	8x6	8x10	9x9	10x8	10x7	10x3	7x3	5x6	10x10	10x7	8x8	9x7	7x10	8x9	9x7	10x10
10x10	9x7	10x9	5x9	5x9	9x5	8x6	6x8	6x10	6x9	6x9	7x8	10x6	6x8	9x6	10x10	9x9	10x10	8x9
9x8	9x7	6x6	10x4	3x7	9x7	10x9	8x9	8x10	8x8	8x10	7x9	10x8	10x4	5x6	5x5	8x8	7x10	7x10
9x9	10x8	5x7	5x10	3x7	8x8	10x8	8x10	10x8	10x10	9x9	10x7	8x8	8x3	7x6	4x8	9x8	9x9	7x10
9x9	9x9	3x10	6x5	3x9	10x10	8x8	10x8	7x10	8x8	9x7	10x7	7x10	3x9	9x3	3x10	7x9	10x10	8x9
10x6	6x8	8x6	7x6	8x7	6x10	6x7	7x8	9x6	8x7	6x10	8x6	7x7	8x7	8x6	6x7	9x5	10x5	7x7

Key:

0	Skin
1-20	Red
21-40	Black
41-60	Gray
61-100	Blue

*Blank squares are dark blue

Name: Date:

Basic Multiplication

Flowers

8x8	1x6	2x4	2x8	1x5	1x6	3x1	4x1	3x3	4x2	7x1	3x2	2x3	4x5	1x4	8x2	8x1	5x1	2x9
10x7	7x9	1x6	1x2	5x10	9x6	6x8	6x2	2x10	1x4	4x3	9x2	1x9	4x1	1x5	10x1	5x4	2x2	10x1
7x9	10x9	8x10	7x9	7x9	10x7	8x8	10x9	4x1	4x2	6x3	1x4	9x1	7x7	10x5	6x9	2x8	10x1	10x1
10x9	7x9	8x8	8x10	8x8	10x10	10x10	9x9	8x10	8x6	5x9	9x6	8x6	8x7	6x8	10x5	6x10	9x7	10x8
8x10	10x9	10x7	10x7	10x10	8x9	8x9	9x9	9x8	7x8	6x9	7x8	10x5	10x6	6x8	7x8	9x6	9x10	9x7
10x8	7x10	7x10	9x9	8x8	7x10	9x7	8x10	10x8	6x7	10x6	6x9	6x10	10x5	9x6	9x5	8x6	7x9	10x10
9x10	10x9	9x10	9x10	8x8	9x9	8x8	10x7	8x9	6x10	7x8	9x6	6x7	8x6	8x6	6x9	7x8	9x10	8x10
7x9	9x7	8x8	9x8	9x7	7x9	9x10	8x8	9x7	7x8	6x10	6x9	9x5	7x7	10x4	5x10	7x6	10x9	7x10
9x9	7x10	9x7	9x7	2x0	8x8	8x9	9x8	9x10	7x7	5x9	6x9	5x10	3x10	7x3	5x8	5x10	10x8	7x9
10x7	10x7	9x9	0x10	8x0	0x8	10x7	8x8	10x7	7x6	6x7	9x6	7x6	5x8	8x3	9x3	5x9	7x10	9x8
8x8	10x9	8x9	0x6	0x0	0x10	9x7	7x9	8x10	5x9	6x8	4x9	10x3	3x10	8x0	3x10	9x4	5x5	8x10
9x8	0x9	1x0	2x0	5x7	0x5	0x2	0x0	8x10	9x6	3x7	8x4	6x6	0x6	4x0	0x6	3x7	3x7	9x4
0x8	0x4	9x0	9x3	7x4	4x6	3x0	9x0	5x0	7x7	7x6	8x5	4x7	9x3	0x0	5x6	7x5	10x3	10x9
7x10	0x3	6x0	0x4	6x6	0x4	0x6	3x0	10x9	7x6	5x10	6x10	2x5	4x10	8x5	5x5	7x8	10x9	10x7
10x8	9x7	7x10	9x0	0x0	0x6	2x5	10x10	10x9	9x6	7x8	3x1	10x5	5x6	5x8	7x4	9x6	9x8	7x10
9x7	8x9	10x8	0x10	0x8	10x0	10x10	2x9	10x10	9x5	9x6	4x2	5x9	9x6	7x3	6x9	6x8	9x8	9x7
9x9	9x10	9x10	8x10	0x1	10x7	7x9	10x9	4x4	10x5	6x7	10x1	6x9	6x7	9x5	5x9	9x6	9x10	8x9
10x9	8x10	7x10	10x7	9x10	9x8	10x8	5x9	2x10	9x6	8x6	7x1	6x8	6x10	10x5	10x5	8x6	8x6	7x10
10x7	9x10	7x9	9x7	7x9	8x8	8x7	6x9	2x6	10x6	8x7	2x9	9x6	7x7	8x7	9x6	7x7	8x7	10x7
5x3	3x1	3x1	3x1	3x4	3x1	2x3	5x2	3x2	5x4	1x8	5x2	2x8	2x9	8x2	4x5	9x2	4x4	2x8

Key:

0	Red
1-20	Green
21-40	Yellow
41-60	Brown
61-100	Blue

Name: Date:

Dividing with 1

Cat

8÷1	10÷1	8÷1	8÷1	8÷1	7÷1	7÷1	10÷1	7÷1	8÷1
10÷1	5÷1	9÷1	7÷1	7÷1	8÷1	10÷1	6÷1	9÷1	8÷1
9÷1	5÷1	5÷1	10÷1	9÷1	10÷1	6÷1	5÷1	10÷1	7÷1
9÷1	5÷1	5÷1	5÷1	5÷1	6÷1	6÷1	6÷1	9÷1	8÷1
9÷1	6÷1		9÷1	5÷1	10÷1		5÷1	10÷1	8÷1
10÷1	6÷1	6÷1	5÷1	6÷1	5÷1	5÷1	5÷1	10÷1	8÷1
10÷1	5÷1	5÷1	6÷1	2÷1	6÷1	6÷1	6÷1	10÷1	7÷1
8÷1	9÷1	5÷1	6÷1	5÷1	5÷1	5÷1	9÷1	7÷1	8÷1
3÷1	4÷1	10÷1	10÷1	9÷1	10÷1	10÷1	3÷1	4÷1	4÷1
4÷1	10÷1	5÷1	5÷1	5÷1	6÷1	5÷1	10÷1	4÷1	3÷1

Key:

1 or 2	Pink
3 or 4	Green
5 or 6	Brown
7 or 8	Blue
9 or 10	Black

*Blank squares are white

Name: Date:

Dividing with 3 Wizard

30÷3	30÷3	30÷3	27÷3	30÷3	30÷3	30÷3	27÷3	30÷3	30÷3
30÷3	27÷3	27÷3	18÷3	18÷3	30÷3	27÷3	30÷3	27÷3	27÷3
30÷3	27÷3	27÷3	30÷3	18÷3	18÷3	27÷3	30÷3	27÷3	30÷3
27÷3		30÷3	15÷3	18÷3	15÷3	15÷3	27÷3	27÷3	30÷3
30÷3	12÷3	30÷3	27÷3	3÷3	6÷3	30÷3	30÷3	27÷3	30÷3
27÷3	3÷3	18÷3	18÷3			15÷3	18÷3	3÷3	27÷3
30÷3	9÷3	18÷3	15÷3		15÷3	15÷3	18÷3	27÷3	30÷3
27÷3	12÷3	27÷3	15÷3	15÷3	15÷3	18÷3	27÷3	30÷3	30÷3
24÷3	9÷3	24÷3	18÷3	15÷3	15÷3	18÷3	21÷3	24÷3	21÷3
21÷3	21÷3	18÷3	15÷3	15÷3	15÷3	18÷3	15÷3	24÷3	21÷3

Key:

1 or 2	Tan
3 or 4	Brown
5 or 6	Gray
7 or 8	Green
9 or 10	Blue

*Blank squares are white

Name: Date:

Dividing with 5 Skateboarding

45÷5	50÷5	50÷5	50÷5	50÷5	50÷5	50÷5	50÷5	50÷5	50÷5
45÷5	45÷5	50÷5	50÷5	35÷5	40÷5	50÷5	50÷5	45÷5	45÷5
50÷5	45÷5	50÷5	45÷5	25÷5	25÷5	50÷5	45÷5	45÷5	45÷5
50÷5	45÷5	45÷5	25÷5	20÷5	20÷5	25÷5	50÷5	50÷5	45÷5
50÷5	45÷5	50÷5	50÷5	15÷5	15÷5	50÷5	50÷5	50÷5	50÷5
50÷5	45÷5	50÷5	45÷5	5÷5	5÷5	50÷5	50÷5	50÷5	45÷5
50÷5	45÷5	50÷5	45÷5	5÷5	10÷5	50÷5	50÷5	45÷5	50÷5
50÷5	50÷5	35÷5	35÷5	40÷5	35÷5	40÷5	35÷5	45÷5	50÷5
50÷5	45÷5	50÷5	10÷5	50÷5	50÷5	10÷5	45÷5	50÷5	50÷5
35÷5	35÷5	35÷5	40÷5	40÷5	35÷5	40÷5	35÷5	35÷5	35÷5

Key:

1 or 2	Black
3 or 4	Red
5 or 6	Tan
7 or 8	Brown
9 or 10	Blue

Name: Date:

Dividing with 7

Campfire

56÷7	56÷7	56÷7	56÷7	63÷7	63÷7	56÷7	63÷7	70÷7	56÷7
49÷7	63÷7	56÷7	49÷7	70÷7	70÷7	56÷7	49÷7	63÷7	56÷7
49÷7	70÷7	63÷7	56÷7	49÷7	56÷7	56÷7	49÷7	56÷7	56÷7
56÷7	63÷7	35÷7	56÷7	42÷7	63÷7	49÷7	56÷7	70÷7	56÷7
63÷7	21÷7	70÷7	56÷7	63÷7	21÷7	70÷7	49÷7	21÷7	63÷7
56÷7	70÷7	28÷7	28÷7	70÷7	21÷7	42÷7	70÷7	21÷7	70÷7
56÷7	21÷7	42÷7	28÷7	21÷7	42÷7	28÷7	28÷7	42÷7	49÷7
56÷7	56÷7	28÷7	35÷7	28÷7	42÷7	28÷7	35÷7	28÷7	49÷7
49÷7	49÷7	21÷7	42÷7	7÷7	7÷7	42÷7	42÷7	28÷7	49÷7
56÷7	7÷7	7÷7	7÷7	21÷7	14÷7	7÷7	7÷7	7÷7	49÷7

Key:

1 or 2	Brown
3 or 4	Red
5 or 6	Yellow
7 or 8	Blue
9 or 10	Gray

Name: Date:

Dividing with 9

Sailboat

18÷9	9÷9	9÷9	9÷9	18÷9	90÷9	81÷9	54÷9	45÷9	45÷9
9÷9	18÷9	18÷9	18÷9	18÷9	9÷9	90÷9	90÷9	45÷9	54÷9
9÷9	18÷9	9÷9	9÷9	18÷9	18÷9	81÷9	90÷9	81÷9	54÷9
9÷9	9÷9	18÷9	9÷9	9÷9	18÷9	81÷9	90÷9	81÷9	90÷9
81÷9	81÷9	27÷9	81÷9	81÷9	90÷9	90÷9	81÷9	81÷9	90÷9
81÷9	90÷9	27÷9	90÷9	90÷9	81÷9	27÷9	27÷9	81÷9	81÷9
36÷9	27÷9	36÷9	36÷9	27÷9	36÷9	27÷9	81÷9	81÷9	90÷9
27÷9	27÷9	27÷9	27÷9	27÷9	27÷9	63÷9	63÷9	63÷9	63÷9
36÷9	36÷9	27÷9	27÷9	27÷9	72÷9	72÷9	72÷9	63÷9	63÷9
63÷9	63÷9	72÷9	63÷9	63÷9	72÷9	72÷9	72÷9	72÷9	63÷9

Key:

1 or 2	Gray
3 or 4	Brown
5 or 6	Yellow
7 or 8	Dark Blue
9 or 10	Blue

Name: Date:

Basic Division Baseball

24÷4	30÷5	12÷2	6÷1	12÷2	12÷2	18÷3	12÷2	24÷4	6÷1	24÷4	18÷3	30÷5	12÷2	24÷4	18÷3	18÷3	24÷4	12÷2
12÷2	18÷3	18÷3	18÷3	18÷3	18÷3	18÷3	18÷3	12÷2	12÷2	6÷1	6÷1	6÷1	12÷2	24÷4	30÷5	12÷2	30÷5	24÷4
30÷5	30÷5	18÷3	6÷1	30÷5	30÷5	18÷3	18÷3	24÷4	6÷1	30÷5	18÷3	30÷5	30÷5	6÷1	12÷2	24÷4	30÷5	6÷1
24÷4	18÷3	30÷5	12÷2	30÷5	6÷1	6÷1	6÷1	18÷3	18÷3	18÷3	30÷5	30÷5	30÷5	18÷3	30÷5	24÷4	30÷5	24÷4
6÷1	24÷4	12÷2	12÷2	12÷2	30÷5	30÷5	12÷2	12÷2	30÷5	30÷5	6÷1	6÷1	18÷3	30÷5	12÷2	18÷3	24÷4	6÷1
30÷5	12÷2	24÷4	12÷2	6÷1	6÷1	12÷2			6÷1	30÷5	24÷4	18÷3	6÷1	18÷3	24÷4	30÷5	12÷2	6÷1
12÷2	18÷3	30÷5	30÷5	12÷2	12÷2	30÷5			6÷1	12÷2	24÷4	18÷3	6÷1	9÷3	6÷2	6÷2	15÷5	30÷5
24÷4	18÷3	24÷4	15÷5	6÷2	9÷3	12÷2	6÷1	24÷4	6÷1	30÷5	12÷2	30÷5	12÷4	15÷5	15÷5	15÷5	12÷4	6÷1
24÷4	12÷2	30÷5	12÷4	15÷5	9÷3	15÷5	30÷5	12÷2	12÷2	18÷3	30÷5	18÷3	18÷3	6÷3	2÷1	4÷2	10÷5	24÷4
24÷4	12÷2	12÷2	2÷1	2÷1	2÷1	10÷5	6÷1	24÷4	30÷5	25÷5	15÷3	18÷3	6÷1	4÷2	10÷5	4÷2	4÷2	12÷2
6÷1	24÷4	18÷3	2÷1	2÷1	6÷3	2÷1	24÷4	24÷4	30÷5	5÷1	20÷4	6÷3	15÷5	3÷1	8÷4	6÷3	4÷2	24÷4
16÷4	8÷2	20÷5	2÷1	6÷3	8÷4	16÷4	4÷1				8÷2	8÷2	20÷5	12÷3	9÷3	6÷2	3÷1	16÷4
12÷3	12÷3	2÷1	6÷1	18÷3	30÷5	12÷3	4÷1			16÷4		4÷2	2÷1	8÷4	12÷4	6÷2	9÷3	16÷4
16÷4	4÷1	10÷5	6÷1	12÷2	6÷3	8÷2	16÷4	16÷4	8÷2	20÷5	4÷1		20÷5	16÷4	6÷2	9÷3	12÷4	4÷1
12÷3	16÷4	8÷4	12÷2	30÷5	30÷5	4÷2	2÷1	8÷2	20÷5	4÷1	4÷1	20÷5		8÷2	20÷5	18÷3	30÷5	8÷2
8÷2	8÷2	16÷4	6÷2	12÷4	15÷5	12÷3	4÷1	4÷1	8÷2	20÷5	4÷1	8÷2	4÷1		20÷5	18÷3	6÷1	16÷4
4÷1	16÷4	9÷3	9÷3	8÷2	12÷4	6÷2	4÷1	16÷4	8÷2	4÷1	20÷5	12÷3				6÷1	30÷5	12÷3
16÷4	15÷5	6÷2	20÷5	12÷3	15÷5	12÷4	12÷3	12÷3	4÷1	16÷4	12÷3				4÷4	3÷3	4÷4	12÷3
	5÷5	5÷5			1÷1	2÷2											20÷5	12÷3
16÷4	4÷1	20÷5	20÷5	8÷2	8÷2	20÷5	16÷4	4÷1	20÷5	16÷4	4÷1	4÷1	4÷1	16÷4	8÷2	4÷1	8:2	4:1

Key:

0	Black
1	Skin
2	Red
3	Green
4	Orange
5	Blue

*Blank squares are white

Name: Date:

Basic Division

Lioness

5÷1	20÷4	24÷8	6÷2	9÷3	20÷4	15÷3	50÷10	15÷3	25÷5	15÷3	25÷5	35÷7	40÷8	9÷3	21÷7	3÷1	20÷4	10÷2
40÷8	24÷8	12÷3	4÷1	28÷7	27÷9	30÷6	50÷10	35÷7	15÷3	40÷8	40÷8	40÷8	30÷10	8÷2	16÷4	36÷9	24÷8	10÷2
24÷8	36÷9	4÷1	18÷9	24÷6	4÷1	12÷4	10÷2	15÷3	40÷8	40÷8	20÷4	12÷4	16÷4	8÷2	20÷10	32÷8	8÷2	21÷7
27÷9	28÷7	2÷1	10÷5	14÷7	32÷8	15÷5	5÷1	45÷9	50÷10	5÷1	45÷9	3÷1	20÷5	2÷1	4÷2	2÷1	20÷5	24÷8
15÷5	12÷3	8÷4	20÷10	8÷4	36÷9	27÷9	3÷1	6÷2	27÷9	3÷1	3÷1	21÷7	40÷10	4÷2	10÷5	14÷7	20÷5	18÷6
3÷1	16÷4	20÷5	16÷4	32÷8	12÷3	32÷8	20÷5	24÷6	32÷8	36÷9	20÷5	20÷5	4÷1	4÷1	40÷10	20÷5	40÷10	15÷5
40÷8	12÷4	36÷9	4÷1	20÷5			20÷5	28÷7	36÷9	28÷7	28÷7			40÷10	32÷8	8÷2	9÷3	20÷4
15÷3	12÷4	40÷10	16÷4	16÷4		8÷4	4÷1	12÷3	20÷5	8÷2	8÷2	8÷4		8÷2	8÷2	24÷6	3÷1	20÷4
20÷4	18÷6	28÷7	12÷3	4÷1		10÷5	20÷5	28÷7	36÷9	4÷1	32÷8	4÷2		8÷2	24÷6	40÷10	6÷2	30÷6
50÷10	27÷9	12÷3	4÷1	28÷7	36÷9	12÷3	28÷7	24÷6	24÷6	4÷1	8÷2	8÷2	24÷6	12÷3	12÷3	20÷5	21÷7	15÷3
50÷10	15÷5	40÷10	4÷1	32÷8	20÷5	32÷8	10÷5	18÷9	18÷9	8÷4	20÷10	24÷6	24÷6	24÷6	4÷1	4÷1	9÷3	20÷4
5÷1	30÷6	15÷5	4÷1	28÷7	40÷10	20÷5	8÷2	4÷2	10÷5	4÷2	32÷8	20÷5	20÷5	28÷7	32÷8	3÷1	25÷5	45÷9
40÷8	50÷10	10÷2	3÷1	40÷10	32÷8	36÷9	4÷4	2÷2	3÷1	7÷7	9÷9	32÷8	40÷10	40÷10	3÷1	10÷2	10÷2	15÷3
25÷5	35÷7	40÷8	35÷7	15÷5	36÷9	24÷8	6÷6	8÷8	21÷7	8÷8	4÷4	18÷6	24÷6	6÷2	15÷3	15÷3	25÷5	15÷3
25÷5	50÷10	24÷8	3÷1	27÷9	8÷2	10÷10	27÷9	9÷3	18÷6	15÷5	9÷3	4÷4	12÷3	27÷9	12÷4	10÷2	30÷6	15÷3
45÷9	6÷2	8÷2	4÷1	9÷3	20÷5	5÷5	1÷1	4÷4	3÷3	9÷9	8÷8	10÷10	12÷3	27÷9	8÷2	27÷9	45÷9	20÷4
9÷3	32÷8	12÷3	20÷5	4÷1	27÷9	18÷6	27÷9	18÷6	3÷1	3÷1	3÷1	18÷6	9÷3	8÷2	12÷3	4÷1	15÷5	20÷4
3÷1	20÷5	12÷3	12÷3	24÷6	10÷10	5÷5	5÷5	6÷6	8÷8	2÷2	2÷2	8÷8	9÷9	12÷3	40÷10	20÷5	24÷6	9÷3
9÷3	36÷9	16÷4	20÷5	4÷4	9÷9	10÷10	9÷9	9÷9	10÷10	3÷3	6÷6	5÷5	10÷10	7÷7	24÷6	28÷7	8÷2	18÷6
30÷10	4÷1	40÷10	40÷10	8÷8	9÷9	8÷8	3÷3	4÷4	1÷1	10÷10	9÷9	10÷10	6÷6	2÷2	4÷1	12÷3	36÷9	15÷5

Key:

1	Tan
2	Black
3	Brown
4	Orange
5	Blue

*Blank squares are white

Name: Date:

Basic Division

Space

20÷4	35÷7	45÷9	20÷10	10÷5	14÷7	6÷3	20÷4	25÷5	20÷4	25÷5	10÷2	15÷3	25÷5	35÷7	25÷5	5÷1	25÷5	15÷3
30÷6	30÷6	24÷8	3÷1	12÷6	12÷6	6÷3	14÷7	15÷3	5÷1	5÷1	30÷6	20÷4	40÷8	10÷2	30÷6	20÷4		35÷7
30÷6	20÷10	3÷1	12÷4	3÷1	20÷10	20÷10	24÷8	18÷6	15÷3	40÷8	15÷3	50÷10	10÷2	50÷10	45÷9	45÷9	25÷5	20÷4
35÷7	20÷10	15÷5	6÷2	15÷5	12÷6	6÷2	4÷2	8÷4	15÷3	15÷3		15÷3	10÷2	10÷2		10÷2	35÷7	10÷2
5÷1	12÷6	2÷1	24÷8	30÷10	6÷3	6÷3	4÷2	10÷5	35÷7	25÷5	20÷4	45÷9	15÷3	50÷10	25÷5	5÷1	25÷5	45÷9
5÷1	10÷5	12÷6	18÷6	8÷4	2÷1	12÷6	6÷2	9÷3	30÷6	30÷6	25÷5	40÷8	50÷10	35÷7	25÷5	20÷4	15÷3	5÷1
25÷5	20÷4	16÷8	14÷7	16÷8	20÷10	18÷6	27÷9	25÷5	20÷4		25÷5	20÷4	30÷6	35÷7	35÷7		40÷8	5÷1
45÷9	15÷3	15÷3	6÷3	18÷9	10÷5	8÷4	35÷7	25÷5	50÷10	5÷1	30÷6	30÷6	15÷3	45÷9	5÷1	50÷10	15÷3	50÷10
	35÷7	45÷9	20÷4	10÷2	50÷10	35÷7	30÷6	5÷1	30÷6	30÷6	25÷5		45÷9	10÷2	50÷10	10÷2	40÷8	45÷9
40÷8	40÷8	50÷10	5÷1	30÷6	45÷9	40÷8		40÷8	25÷5	5÷1	45÷9	15÷3	50÷10	10÷2	25÷5	20÷4	15÷3	45÷9
10÷2	5÷1	40÷8		15÷3	40÷8	35÷7	50÷10	15÷3	40÷8	40÷8	50÷10	40÷8	5÷1	25÷5	40÷8	10÷2	40÷8	5÷1
10÷2	35÷7	35÷7	30÷6	25÷5	40÷8	50÷10	25÷5	15÷3	35÷7		15÷3	45÷9	20÷4	30÷6	25÷5	40÷8	45÷9	35÷7
40÷8	15÷3	20÷4	45÷9	35÷7	10÷2	50÷10	45÷9	15÷3	10÷2	45÷9	50÷10	50÷10	5÷1		30÷6	35÷7	50÷10	30÷6
30÷6	30÷6	30÷6	50÷10		25÷5	50÷10	45÷9	20÷4	20÷4	25÷5	10÷2	35÷7	50÷10	5÷1	30÷6	20÷4	30÷6	35÷7
35÷7	30÷6	5÷1	50÷10	50÷10	25÷5	15÷3	25÷5	40÷8		20÷4	25÷5	35÷7	50÷10	35÷7	25÷5	20÷4	25÷5	10÷2
35÷7	30÷6	5÷1	25÷5	5÷1	20÷4	10÷2	5÷1	35÷7	40÷8	15÷3	10÷2	15÷3	45÷9	40÷8	35÷7	20÷4		35÷7
25÷5		25÷5	30÷6	50÷10	50÷10	15÷3	15÷3	15÷3	15÷3	45÷9	5÷1	5÷1	50÷10	15÷3	50÷10	25÷5	5÷1	15÷3
20÷4	40÷8	5÷1	50÷10	20÷4	45÷9	40÷8	10÷2	10÷2	20÷4	50÷10	50÷10	20÷4	15÷3	45÷9	45÷9	50÷10	30÷6	35÷7
35÷7	25÷5	25÷5	35÷7	12÷3	16÷4	4÷1	40÷10	12÷3	12÷3	12÷3	28÷7	4÷1	4÷1	32÷8	25÷5	10÷2	40÷8	40÷8
40÷8	40÷10	28÷7	32÷8	4÷1	36÷9	28÷7	16÷4	12÷3	4÷1	36÷9	12÷3	36÷9	16÷4	12÷3	4÷1	36÷9	20÷5	10÷2

Key:

1	Blue
2	Green
3	Gray
4	Black

*Blank squares are white

Name: Date:

Advanced Division

Cherries

63÷9	28÷4	72÷9	24÷3	40÷5	35÷5	35÷5	49÷7	70÷10	80÷10	63÷9	35÷5	48÷6	80÷10	8÷1	63÷9	16÷2	35÷5	32÷4
56÷7	40÷5	35÷5	8÷1	28÷4	48÷6	56÷8	56÷8	6÷6	4÷2	8÷1	21÷3	49÷7	48÷6	10÷5	10÷5	72÷9	80÷10	42÷6
40÷5	56÷7	42÷6	28÷4	64÷8	49÷7	56÷8	40÷5	24÷3	32÷4	10÷10	56÷7	72÷9	20÷10	6÷1	10÷2	2÷1	7÷1	63÷9
35÷5	49÷7	49÷7	48÷6	80÷10	16÷2	80÷10	72÷9	64÷8	21÷3	5÷5	10÷10	4÷2	42÷7	24÷4	54÷9	35÷7	1÷1	40÷5
7÷1	21÷3	28÷4	56÷7	49÷7	72÷9	14÷2	56÷8	35÷5	4÷2	32÷4	7÷7	16÷8	3÷3	8÷4	4÷2	4÷4	10÷10	16÷2
63÷9	21÷3	42÷6	64÷8	10÷10	2÷2	80÷10	56÷8	6÷6	72÷9	8÷1	21÷3	10÷5	72÷9	35÷5	49÷7	32÷4	28÷4	40÷5
21÷3	56÷7	14÷2	8÷4	6÷2	12÷4	6÷3	5÷5	56÷8	64÷8	28÷4	56÷8	18÷9	7÷1	70÷10	49÷7	70÷10	35÷5	24÷3
80÷10	16÷2	1÷1	12÷4	24÷8	32÷8	15÷5	4÷2	48÷6	72÷9	21÷3	40÷5	8÷1	4÷2	40÷5	64÷8	21÷3	56÷7	42÷6
56÷7	12÷6	27÷9	9÷3	15÷5	21÷7	4÷1	4÷1	10÷10	56÷8	70÷10	8÷1	56÷7	7÷7	21÷3	14÷2	8÷1	32÷4	21÷3
35÷5	7÷7	18÷6	32÷8	16÷4	18÷6		40÷10	14÷7	56÷8	24÷3	56÷8	16÷2	56÷7	3÷3	14÷2	49÷7	48÷6	32÷4
48÷6	21÷3	6÷6	6÷2	36÷9		9÷3	9÷9	63÷9	16÷2	56÷7	7÷1	32÷4	6÷3	18÷9	8÷8	72÷9	32÷4	8÷1
72÷9	40÷5	28÷4	1÷1	6÷2	3÷1	4÷4	32÷4	24÷3	14÷2	35÷5	48÷6	63÷9	8÷8	15÷5	16÷4	7÷7	35÷5	16÷2
56÷8	40÷5	48÷6	40÷5	8÷8	2÷1	49÷7	7÷1	56÷7	40÷5	49÷7	70÷10	8÷4	12÷3	18÷6	20÷5	15÷5	10÷5	42÷6
42÷6	56÷7	16÷2	72÷9	49÷7	7÷1	56÷8	35÷5	48÷6	35÷5	24÷3	8÷8	21÷7	24÷8	30÷10	28÷7	18÷6	24÷6	14÷7
48÷6	49÷7	48÷6	48÷6	32÷4	14÷2	28÷4	32÷4	14÷2	48÷6	49÷7	14÷7	8÷2	15÷5	27÷9	15÷5		12÷3	14÷7
35÷5	24÷3	56÷8	70÷10	56÷7	49÷7	24÷3	32÷4	63÷9	24÷3	40÷5	56÷8	4÷2	24÷8	30÷10		27÷9	4÷4	63÷9
63÷9	7÷1	49÷7	64÷8	48÷6	32÷4	56÷8	16÷2	7÷1	48÷6	32÷4	56÷7	56÷8	9÷9	4÷1	30÷10	7÷7	24÷3	56÷7
24÷3	42÷6	42÷6	32÷4	21÷3	80÷10	16÷2	49÷7	56÷8	63÷9	16÷2	7÷1	28÷4	14÷2	20÷10	1÷1	64÷8	8÷1	7÷1
28÷4	14÷2	28÷4	49÷7	35÷5	72÷9	80÷10	8÷1	14÷2	35÷5	48÷6	63÷9	48÷6	64÷8	21÷3	8÷1	24÷3	64÷8	56÷8
56÷8	32÷4	56÷8	49÷7	21÷3	40÷5	32÷4	24÷3	70÷10	42÷6	32÷4	42÷6	40÷5	24÷3	49÷7	72÷9	56÷8	48÷6	7÷1

Key:

1,2	Black
3,4	Red
5,6	Green
7,8	Blue

*Blank squares are white

Name: Date:

Advanced Division

Pig

56÷8	8÷1	3÷3	4÷2	4÷2	70÷10	56÷7	8÷4	20÷10	8÷8	2÷1	40÷5	14÷2	4÷2	12÷6	1÷1	80÷10	56÷8	35÷5
56÷7	16÷2	10÷5	30÷10	24÷6	2÷2	10÷10	15÷5	15÷5	28÷7	21÷7	10÷10	3÷3	3÷1	12÷4	12÷6	80÷10	32÷4	70÷10
24÷3	70÷10	8÷4	6÷2	8÷8	27÷9	24÷6	24÷6	12÷3	21÷7	21÷7	9÷3	18÷6	6÷3	18÷6	5÷5	64÷8	7÷1	70÷10
49÷7	48÷6	70÷10	4÷2	32÷8	6÷2		9÷3	30÷10	15÷5	28÷7		28÷7	4÷1	8÷4	28÷4	32÷4	24÷3	7÷1
56÷8	56÷7	7÷7	12÷3	36÷9	27÷9	8÷8	4÷1	32÷8	24÷8	16÷4	4÷4	15÷5	30÷10	27÷9	3÷3	32÷4	80÷10	32÷4
32÷4	24÷3	2÷1	36÷9	32÷8	12÷3	12÷4	9÷3	21÷7	6÷2	30÷10	9÷3	6÷2	15÷5	32÷8	2÷2	28÷4	48÷6	49÷7
8÷1	7÷1	10÷5	24÷8	6÷2	16÷4	24÷6	40÷10	6÷2	8÷2	36÷9	4÷1	21÷7	20÷5	9÷3	1÷1	48÷6	7÷1	35÷5
28÷4	7÷7	36÷9	8÷2	24÷8	3÷1	8÷2	14÷7	5÷5	8÷8	8÷4	20÷10	4÷4	12÷3	4÷1	9÷3	8÷4	48÷6	7÷1
7÷7	20÷5	16÷4	6÷2	20÷5	24÷6	12÷6	40÷10	28÷7	16÷4	3÷1	18÷6	15÷5	3÷3	30÷10	40÷10	6÷3	48÷6	2÷1
20÷10	18÷6	32÷8	21÷7	16÷4	20÷5	18÷9	24÷8	10÷10	20÷5	8÷2	2÷2	28÷7	2÷1	24÷6	16÷4	1÷1	18÷9	21÷7
6÷3	6÷2	40÷10	24÷6	16÷4	21÷7	20÷10	6÷2	12÷4	12÷3	12÷3	21÷7	16÷4	1÷1	4÷1	21÷7	3÷3	36÷9	5÷5
5÷5	3÷1	18÷6	21÷7	28÷7	40÷10	15÷5	14÷7	4÷2	10÷10	5÷5	14÷7	3÷3	20÷5	40÷10	4÷1	18÷9	4÷2	48÷6
16÷8	40÷10	6÷2	9÷3	28÷7	32÷8	12÷4	18÷6	16÷8	16÷4	6÷2	10÷10	24÷8	8÷2	16÷4	16÷4	18÷9	56÷7	7÷1
5÷5	12÷3	40÷10	18÷6	12÷3	6÷2	30÷10	32÷8	40÷10	20÷10	12÷6	15÷5	27÷9	20÷5	24÷6	12÷3	4÷4	14÷2	42÷6
2÷2	6÷2	18÷6	9÷3	3÷1	16÷4	28÷7	3÷1	40÷10	9÷3	20÷5	28÷7	24÷8	8÷2	6÷2	10÷10	35÷7	50÷10	48÷8
30÷5	18÷9	20÷5	8÷2	27÷9	20÷5	36÷9	3÷1	21÷7	20÷5	15÷5	9÷3	3÷1	9÷3	10÷10	10÷2	30÷5	24÷4	35÷7
25÷5	30÷6	9÷9	15÷5	1÷1	2÷2	12÷4	27÷9	9÷3	8÷2	8÷4	6÷3	4÷1	6÷3	35÷7	36÷6	42÷7	18÷3	45÷9
30÷6	5÷1	30÷6	4÷2	8÷8	20÷10	6÷6	10÷10	14÷7	12÷6	1÷1	16÷8	1÷1	30÷5	35÷7	5÷1	54÷9	42÷7	30÷6
12÷2	40÷8	24÷4	1÷1	10÷5	20÷10	2÷1	25÷5	48÷8	10÷10	20÷10	7÷7	10÷10	50÷10	20÷4	15÷3	25÷5	50÷10	45÷9
50÷10	24÷4	36÷6	5÷1	60÷10	30÷5	45÷9	5÷1	45÷9	35÷7	30÷6	24÷4	30÷5	18÷3	10÷2	30÷6	10÷2	20÷4	20÷4

Key:

1,2	Black
3,4	Pink
5,6	Brown
7,8	Blue

*Blank squares are white

Regrouping

Name: Date:

Addition with Regrouping to Tens

Baby Chick

17 + 15	13 + 32	34 + 14	27 + 13	18 + 28	34 + 14	20 + 29	21 + 18	22 + 20	24 + 26
15 + 11	20 + 18	35 + 15	17 + 39	34 + 19	33 + 27	18 + 39	15 + 33	17 + 22	16 + 22
25 + 13	13 + 31	29 + 42	48 + 19	38 + 21	16 + 46	32 + 34	33 + 19	31 + 17	13 + 36
27 + 22	47 + 13	45 + 23		29 + 39	44 + 30		15 + 43	49 + 19	28 + 12
18 + 23	22 + 31	35 + 24	13 + 46	37 + 28	32 + 43	40 + 14	43 + 11	37 + 29	24 + 12
16 + 14	22 + 46	29 + 31	35 + 36	44 + 34	48 + 49	26 + 32	21 + 43	34 + 31	36 + 11
12 + 14	30 + 37	11 + 45	24 + 39	21 + 49	43 + 26	19 + 50	41 + 23	49 + 25	33 + 14
25 + 21	34 + 30	41 + 23	37 + 15	21 + 46	35 + 27	30 + 21	18 + 47	24 + 34	26 + 19
16 + 24	19 + 16	23 + 36	30 + 24	23 + 28	25 + 48	18 + 34	50 + 12	19 + 15	25 + 16
31 + 12	29 + 16	41 + 43	32 + 47	23 + 22	19 + 13	48 + 39	50 + 47	16 + 24	11 + 28

Key:

From 26 to 50	Blue
From 51 to 75	Yellow
From 76 to 100	Orange

*Blank squares are black

Name: Date:

Addition with Regrouping to Hundreds

Cowboy

252 + 575	250 + 549	422 + 450	220 + 682	302 + 693	177 + 244	220 + 557	364 + 419	469 + 437	109 + 834
566 + 386	720 + 139	180 + 196	649 + 265	191 + 182	245 + 156	214 + 233	637 + 259	218 + 134	423 + 526
299 + 644	326 + 264	233 + 210	131 + 214	163 + 129	166 + 191	225 + 220	214 + 223	154 + 237	548 + 411
422 + 491	610 + 504	429 + 338	403 + 553	346 + 367	324 + 316	364 + 319	472 + 296	243 + 512	348 + 417
711 + 262	715 + 483	381 + 538	193 + 678	200 + 167	307 + 357	187 + 151	330 + 662	598 + 332	490 + 422
504 + 353	821 + 222	229 + 544	246 + 536	330 + 355	562 + 162	421 + 105	408 + 486	483 + 315	606 + 223
431 + 536	528 + 704	245 + 882	190 + 137	455 + 689	614 + 557	707 + 443	148 + 166	474 + 685	896 + 263
422 + 491	711 + 262	504 + 353	177 + 147	159 + 213	262 + 760	239 + 154	238 + 209	473 + 474	557 + 663
578 + 782	348 + 962	384 + 936	141 + 209	154 + 240	976 + 239	176 + 153	228 + 137	635 + 784	433 + 662
379 + 936	938 + 336	549 + 838	183 + 145	159 + 128	180 + 899	163 + 215	232 + 234	750 + 800	456 + 181

Key:

From 251 to 500	Black
From 501 to 750	Skin
From 751 to 1,000	Light Blue
From 1,001 to 1,250	Dark Blue
From 1,251 to 1,500	Green

Name: Date:

Addition with Regrouping to Hundreds

Truck

240 + 529	525 + 252	480 + 518	278 + 483	812 + 135	469 + 323	270 + 641	582 + 369	520 + 374	127 + 782
591 + 176	607 + 151	189 + 694	697 + 181	444 + 332	459 + 353	276 + 591	325 + 507	647 + 318	385 + 441
348 + 434	717 + 150	638 + 132	698 + 145	870 + 298	549 + 605	658 + 252	470 + 352	164 + 804	480 + 510
194 + 753	592 + 301	278 + 483	702 + 224	736 + 365	540 + 409	861 + 294	144 + 744	646 + 272	748 + 250
575 + 480	680 + 570	951 + 108	799 + 327	888 + 154	325 + 815	734 + 374	154 + 915	434 + 629	289 + 528
563 + 514	187 + 257	188 + 275	902 + 286	218 + 873	737 + 264	290 + 209	148 + 283	801 + 395	535 + 425
176 + 775	206 + 243	383 + 107	118 + 815	339 + 537	514 + 316	229 + 240	328 + 172	489 + 308	702 + 205
131 + 519	605 + 145	510 + 160	213 + 298	473 + 207	329 + 240	519 + 122	268 + 300	264 + 435	143 + 634
209 + 476	168 + 482	461 + 148	297 + 354	245 + 372	311 + 381	332 + 280	200 + 430	115 + 475	148 + 815
285 + 264	407 + 198	505 + 211	265 + 375	351 + 268	382 + 294	233 + 366	433 + 267	570 + 316	150 + 614

Key:

From 251 to 500	Black
From 501 to 750	Brown
From 751 to 1,000	Blue
From 1,001 to 1,250	Red

Name: Date:

Subtraction with Regrouping to Tens

Rabbit

48 - 13	47 - 11	74 - 52	44 - 15	60 - 21	91 - 78	71 - 35	81 - 63	58 - 19	63 - 35
73 - 33	93 - 64	49 - 20	96 - 59	84 - 83	69 - 15	49 - 43	65 - 21	52 - 50	46 - 10
75 - 44	48 - 21	76 - 55	41 - 15	69 - 50	87 - 31	65 - 62	85 - 34	49 - 42	95 - 70
27 - 16	60 - 59	34 - 27	22 - 20	26 - 22	84 - 31	63 - 47	61 - 19	63 - 51	42 - 16
72 - 31	99 - 47	86 - 32	37 - 19	72 - 68	84 - 29	84 - 24	63 - 14	91 - 46	25 - 11
97 - 49	78 - 22	74 - 29	96 - 41	56 - 13	94 - 36	64 - 15	63 - 51	68 - 11	22 - 10
95 - 49	92 - 46	80 - 25	72 - 18	62 - 15	77 - 36	91 - 44	81 - 28	90 - 40	86 - 68
80 - 23	55 - 41	36 - 16	74 - 16	98 - 38	89 - 46	60 - 49	96 - 77	75 - 56	94 - 73
80 - 35	55 - 12	79 - 25	79 - 75	73 - 34	92 - 32	69 - 20	80 - 23	26 - 24	73 - 35
80 - 62	65 - 48	75 - 57	48 - 17	74 - 50	81 - 71	86 - 81	53 - 45	62 - 28	96 - 66

Key:

From 1-20	Black
From 21-40	Green
From 41-60	Gray

Name: Date:

Subtraction with Regrouping to Hundreds — Dolphin

985 - 242	802 - 178	873 - 258	884 - 202	929 - 207	930 - 305	915 - 135	827 - 184	953 - 209	891 - 157
921 - 220	566 - 234	901 - 242	921 - 179	993 - 323	941 - 322	940 - 233	987 - 333	976 - 282	840 - 216
829 - 148	897 - 573	910 - 685	832 - 213	876 - 154	945 - 308	917 - 158	876 - 267	766 - 110	884 - 155
445 - 223	629 - 349	833 - 502	695 - 340	907 - 544	830 - 568	967 - 728	821 - 603	838 - 106	856 - 166
575 - 176	370 - 139	997 - 706	727 - 484	559 - 305	926 - 681	656 - 515	890 - 284	638 - 387	779 - 427
844 - 562	967 - 593	706 - 397	476 - 214	473 263	525 - 126	636 - 242	646 - 282	911 - 345	554 - 145
370 - 139	560 - 352	312 - 102	937 - 395	778 - 252	506 - 192	860 - 501	994 - 709	937 - 327	827 - 183
926 - 176	841 - 132	835 - 196	955 - 553	898 - 318	857 - 191	909 - 229	928 - 164	927 - 309	898 - 226
933 - 224	902 - 125	805 - 108	665 - 184	981 - 283	778 - 124	958 - 246	950 - 167	839 - 110	919 - 254
914 - 165	872 - 160	734 - 110	978 - 347	864 - 192	875 - 229	909 - 229	811 - 120	807 - 156	924 - 172

Key:

From 1 to 200	Black
From 201 to 400	Light Gray
From 401 to 600	Dark Gray
From 601 to 800	Blue

Name: Date:

Subtraction with Regrouping to Hundreds

Zombie

864 - 204	946 - 294	933 - 275	774 - 128	836 - 146	952 - 265	998 - 310	910 - 134	949 - 203	968 - 336
929 - 147	340 - 182	686 - 550	858 - 172	968 - 346	972 - 268	423 - 318	951 - 912	885 - 211	979 - 196
873 - 260	778 - 505	365 - 144	932 - 141	861 - 165	921 - 255	587 - 171	872 - 276	832 - 230	894 - 176
945 - 320	517 - 286	461 - 148	924 - 549	986 - 255	932 - 479	958 - 547	858 - 267	758 - 121	872 - 141
908 - 225	665 - 346	814 - 582	812 - 112	780 - 134	877 - 223	753 - 201	786 - 186	891 - 218	812 - 210
913 - 143	944 - 863	909 - 817	860 - 655	953 - 276	877 - 454			721 - 268	941 - 298
887 - 134	459 - 441	988 - 802	899 - 279	970 - 216	912 - 266			925 - 196	902 - 259
922 - 281			953 - 197	863 - 138	970 - 233	803 - 616	201 - 175	787 - 171	763 - 113
776 - 157			943 - 248	997 - 335	793 - 150	352 - 159	581 - 478	869 - 190	908 - 295
780 - 549	799 - 416	877 - 546	954 - 594	973 - 755	492 - 231	859 - 478	598 - 371	956 - 695	804 - 589

Key:

From 1 to 200	Black
From 201 to 400	Green
From 401 to 600	Skin
From 601 to 800	Blue

*Blank squares are purple

Math Concepts

Name: Date:

Place Value Multiplication

Fun Alligator

4x80	4x80	8x50	4x80	4x90	9x40	6x60	4x90	8x50	7x50	5x70	4x80	8x40	5x80	4x90	8x40	4x90	7x50	8x40
4x90	4x80	4x80	4x90	8x50	5x70	4x80	8x40	4x90	5x70	5x70	7x50	5x80	6x60	8x40	6x60	4x80	4x80	6x60
8x50	5x80	5x80	4x90	4x90	6x50	4x60	7x40	5x70	4x60	3x80	5x80	8x40	8x50	5x70	5x70	5x80	4x80	5x70
5x70	4x90	8x40	6x60	4x70	9x30			5x60			5x50	7x50	8x50	4x80	4x90	4x90	5x70	5x70
8x40	5x80	7x50	3x70	6x40		1x60	1x30	3x70	2x10		7x30	9x40	4x90	7x50	6x60	7x50	4x90	5x80
8x40	5x70	4x90	9x40	3x80		3x30	2x20	4x70	3x30		6x40	8x50	7x50	3x90	3x70	9x30	6x60	4x90
9x40	7x50	6x60	5x50	5x50	4x70			8x30		6x40	9x30	7x30	3x70	1x50	6x50	1x90	9x30	8x40
9x40	4x90	4x90	4x90	7x40	8x30	5x60	3x90	8x30	7x40	3x80	7x30	9x30	9x30	8x30	6x50	7x40	6x40	9x40
5x80	5x70	8x40	9x30	4x70	3x40	4x50	3x50	9x20		2x90		6x20		4x50		6x60	6x60	6x60
5x70	9x40	5x70	5x80	7x40	9x30	2x90	9x20	4x40	9x20	5x30	8x20	8x40	5x70	5x70	9x40	8x50	9x40	4x90
8x40	8x40	8x40	3x90	8x30	5x50	3x40	5x40	3x60	5x30	7x20	6x60	5x70	4x90	6x60	8x40	9x40	5x80	4x80
5x80	8x50	7x50	4x90	6x40	6x40	2x70	8x20	4x50	3x40	6x60	4x70	9x40	5x80	4x80	5x70	5x80	6x60	8x50
5x80	8x50	6x60	5x60	6x50	7x30	6x40	4x60	5x60	6x50	3x70	8x30	3x70	8x50	8x40	8x40	8x50	6x60	5x70
4x80	7x50	6x60	8x40	6x50	5x50	8x30	4x60	4x60	3x90	4x70	3x80	3x90	7x50	8x40	8x40	8x50	7x50	4x90
5x70	4x80	4x80	3x90	8x30	5x60	3x40	2x70	8x20	5x40	6x60	5x80	5x70	4x90	4x90	4x90	4x90	4x90	5x70
8x50	8x50	6x60	5x70	3x90	9x30	5x30	7x20	4x40	7x20	5x30	4x90	8x50	4x90	5x70	8x50	5x70	4x80	5x80
8x50	4x90	5x80	3x80	5x50	6x50	9x30	3x60	4x40	3x60	3x50	5x70	4x90	5x80	6x60	7x50	9x40	9x40	4x80
3x90	5x70	7x40	5x70	6x50	7x40	6x50	5x40	2x80	3x50	8x20	8x30	5x70	5x70	4x90	4x80	9x40	7x50	4x90
5x50	4x70	9x30	3x90	3x90	7x40	5x50	8x30	5x50	2x90	3x70	4x60	8x40	8x40	9x40	8x40	5x80	8x50	6x60
3x70	7x30	6x40	5x60	4x60	5x60	3x70	8x30	5x60	6x60	4x60	6x50	7x50	9x40	4x80	6x60	8x40	8x50	9x40

Key:

1-100	Black
101-200	Light Green
201-300	Dark Green
301-400	Blue

*Blank squares are white

Name: Date:

Place Value Multiplication

Basketball

				7x80	6x90	8x60	8x80	9x70	5x90	8x90	6x90	6x70	7x70	7x80	8x60	8x90	6x90	8x80
					7x60	8x90	9x90	6x70	8x90	8x90	6x70	6x70	6x90	6x90	9x60	8x90	8x90	7x90
					9x50	9x70	7x70	8x90	8x70	6x70	6x80	8x90	6x90	7x80	7x90	7x60	5x90	9x60
		7x40	3x70	7x40	8x60	8x70	6x80	5x90	6x90	8x70	9x80	6x90	8x70	8x60	7x60	5x90	8x90	8x70
	3x80				8x30	5x90	6x70	7x70	7x70	9x50	8x60	5x90	6x80	8x60	9x60	6x70	7x60	8x70
		7x30	7x30	6x40	6x70	8x70	7x80	5x90	9x80	8x60	9x80	6x80	5x20	3x20	1x60	8x10	1x10	7x80
7x60	7x10	6x80	8x90	8x90	6x80	8x60	7x80	9x60	7x70	6x70	9x80	9x50	7x60	2x10	3x10	1x50	1x80	6x70
6x80	7x10	7x80	9x50	6x90	8x90	9x50	7x70	6x80	7x60	8x70	6x90	8x80	6x60	4x90	4x90	4x20	1x50	4x10
9x80	8x10	8x90	5x90	9x70	8x90	6x90	8x60	7x90	5x60	9x30	6x90	9x70	8x40	5x80	8x40	1x60	1x90	9x10
6x70	8x10	9x70	9x80	9x50	7x60	6x90	8x60	3x90	3x80	3x80	4x70	8x50	6x40	7x50	4x90	9x40	1x30	7x10
9x60	3x20	6x90	7x60	9x90	7x80	8x80	9x70	9x30	9x30	7x30	5x60	7x70	8x80	5x60	4x60	3x70	1x50	1x90
8x80	5x20	7x70	5x90	6x80	7x60	6x70	8x70	8x80	8x30	9x30	7x50	7x50	4x80	7x40	9x30	7x30	4x10	7x60
6x70	1x50	7x60	7x80	6x70	9x60	9x80	8x70	7x60	8x60	9x70	8x90	8x90	9x90	3x70	6x40	6x40	5x90	9x60
9x60	4x20	7x70	6x80	6x70	8x60	7x70	8x90	7x80	6x80	9x60	9x50	8x80	6x70	7x60	1x90	1x90	8x90	9x70
2x70	1x50	7x20	7x20	5x40	2x80	6x20	4x50	6x30	4x40	2x70	2x90	8x20	2x70	5x30	3x10	1x50	2x90	5x30
4x50	1x60	3x50													4x80	7x50		
3x40	7x10		2x60	4x30	3x50	3x50	9x20	8x20	2x90	6x30	5x40	2x90	3x50	3x20	1x80	3x30	4x40	4x40
6x30		3x40	6x30	8x20	6x20	8x20	4x40	2x70	8x20	8x20	6x20	3x40	5x30	4x50	3x50	9x20	5x30	2x60
	4x30	4x40	4x50	2x90	4x50	4x30	3x60	7x20	5x40	4x50	4x50	3x40	9x20	8x20	7x20	4x30	4x30	6x20
2x60	4x30	9x20	3x60	2x70	3x50	8x20	3x40	5x40	4x30	6x30	5x30	3x60	5x40	2x70	3x50	3x60	3x40	5x30

Key:

1-100	Black
101-200	Brown
201-300	Orange
301-400	Skin
401-1,000	Blue

*Blank squares are white

Name: Date:

Place Value Multiplication

Castle

9x40	6x60	4x80	5x70	8x40	6x60	4x80	5x70	4x80	6x60	5x70	7x50	4x90	5x70	9x40	8x40	8x40	6x60	8x50
9x40	1x20	4x90	2x20	9x40	5x20	8x40	5x70	7x50	7x50	8x40	6x60	6x60	9x10	7x50	1x40	5x80	2x20	9x40
1x10	7x40	4x10	6x50	1x70	3x90	9x10	8x50	6x60	6x60	4x90	5x70	3x20	3x80	1x70	6x40	5x10	3x70	6x10
1x30	9x30	7x30	3x70	5x50	3x80	1x70	4x90	8x50	4x80	9x40	4x90	3x20	7x30	7x30	3x90	3x80	3x90	2x50
6x60	1x30	2x20	3x20	7x10	5x10	4x90	5x80	9x40	5x80	5x70	6x60	8x40	1x90	2x20	1x90	1x90	1x20	7x50
8x50	4x80	1x80	8x30	3x30	6x60	5x70	7x50	9x40	5x70	8x50	4x90	8x40	4x90	1x50	7x40	1x50	5x80	5x80
6x60	8x50	1x20	8x30	3x20	2x40	5x70	1x90	5x70	4x20	5x70	1x50	4x90	1x10	2x20	7x40	2x10	4x90	8x40
4x90	5x80	9x10	6x40	2x50	3x80	3x20	3x80	1x50	3x90	1x60	7x40	5x10	5x50	3x20	5x50	1x30	4x90	4x80
7x50	4x90	3x30	4x70	1x50	9x30	9x30	3x90	5x60	8x30	4x70	3x70	9x30	7x30	2x50	7x30	3x20	8x50	4x90
4x80	4x90	2x20	3x70	9x10	6x40	8x30	4x60	3x70	9x30	4x60	4x60	3x70	3x90	3x10	3x70	1x10	5x80	6x60
8x50	5x70	1x60	8x30	2x20	9x30	6x50	3x90	6x40	5x50	3x90	4x70	7x40	8x90	8x80	6x50	5x10	7x50	9x40
5x80	9x60	7x90	8x30	1x90	6x50	5x50	2x40	9x10	8x30	3x90	9x30	9x90	9x60	8x90	5x90	2x20	7x50	6x60
7x70	9x70	9x50	6x70	8x10	7x40	2x50	3x10	2x30	1x60	4x70	7x30	7x70	5x90	8x70	8x90	4x10	9x40	8x40
6x70	9x80	9x60	6x90	4x20	5x50	6x10	2x30	5x10	1x50	4x70	9x30	7x70	9x50	7x90	9x50	3x10	6x60	5x80
6x80	7x60	8x60	7x80	1x80	7x30	4x10	1x70	2x40	9x10	3x90	3x70	9x90	6x90	6x80	6x90	3x30	5x80	5x80
7x50	6x30	6x30	3x70	1x40	9x30	1x20	1x30	5x10	4x20	5x50	3x70	3x70	3x40	2x90	5x50	1x70	4x80	8x40
5x10	3x50	3x40	4x10	5x20	4x20	4x10	1x10	8x10	1x40	1x10	1x30	1x50	3x60	3x40	1x70	1x80	9x10	1x90
7x70	7x90	9x60	7x80	7x60	7x70	7x70	8x60	8x70	8x80	5x90	7x80	9x70	3x60	3x60	8x60	7x70	7x80	7x70
9x80	7x90	6x70	5x90	8x80	9x60	7x60	7x90	9x50	9x70	6x80	8x70	9x50	4x50	6x30	9x80	9x90	9x50	9x80
5x90	8x70	8x70	9x50	9x60	9x80	8x60	9x50	6x80	7x80	9x50	8x70	7x70	7x90	6x80	6x80	9x60	6x70	7x90

Key:

1-100	Black
101-200	Brown
201-300	Gray
301-400	Blue
401-1,000	Green

Name: Date:

Rounding to Tens

Fishbowl

29	29	27	27	31	28	34	31	28	29	32	32	30	31	34	34	27	32	27
28	29	33	25	25	33	28	27	29	32	30	33	34	30	25	29	26	27	34
32	26	30	34	29	30	31	34	29	33	34	26	30	27	34	32	31	32	28
30	26	28	6	13	6	6	10	7	6	5	13	10	10	12	14	27	32	31
28	26	33	14	14	7	12	5	12	8	12	8	9	9	14	10	32	30	28
32	32	27	31	5	10	14	7	6	6	9	12	9	8	7	26	27	31	29
32	30	28	7	5	8	14	9	13	9	8	5	13	11	5	9	26	28	28
33	34	14	6	13	12	10	14	10	13	9	5	9	7	10	11	9	25	30
32	28	18	21	15	17	15	22	17	15	20	22	22	22	16	19	20	33	31
31	29	21		16				20	16	21	23	19	16	21	15	19	34	28
28	32	22					41		16	18	22	16	15	17	18	24	32	26
30	29	18							17	23	23	24	19	20	24	16	29	33
26	27	22		24				17	15	41		24		17	17	17	31	26
27	26	22	23	15	16	22	17	23						15	20	24	28	29
32	28	32	19	15	15	24	18	18	22			17		17	16	28	29	26
30	27	28	27	23	19	21	15	22	18	20	24	18	19	24	31	30	30	31
38	37	41	38	36	35	39	41	40	35	43	36	44	38	42	42	38	40	44
39	36	37	43	41	35	37	36	38	43	37	40	35	41	41	43	39	37	44
38	38	36	35	37	36	39	39	35	40	38	35	35	44	42	35	40	40	37
38	35	37	36	35	44	44	41	37	36	44	39	39	42	40	39	40	41	43

Key:

Rounds to 10	Light Blue
Rounds to 20	Dark Blue
Rounds to 30	Gray
Rounds to 40	Black

*Blank squares are orange

Name: Date:

Rounding to Hundreds

Hockey Stick

302	335	259	323	287	320	302	264	280	305	280	335	343	293	310	269	265	314	303
321	273	318	299	349	272	321	313	268	347	282	331	286	299	322	274	335	280	715
281	341	336	278	320	315	285	335	306	271	268	336	317	348	256	312	281	720	683
272	314	272	287	294	312	320	272	323	299	342	293	301	330	299	308	695	663	654
283	330	666	695	736	667	653	680	728	747	690	689	717	747	729	688	703	714	721
279	657	693	651	650	740	733	719	668	653	709	700	658	658	725	673	663	706	652
278	650	692	650	690	722	681	730	737	675	650	722	723	685	747	667	721	717	260
290	726	731	532	548	479	514	479	525	497	481	470	463	663	697	726	742	713	256
348	738	471	530	499	547	513	497	465	541	501	479	485	510	724	650	735	300	306
331	688	465	485	519	490	486	491	469	513	514	524	515	512	674	650	311	318	327
431	421	580	488	488	496	547	484	508	525	454	513	505	560	442	393	422	354	430
419	365	572	613	599	593	640	577	608	573	611	568	577	607	423	387	409	394	408
432	432	404	619	641	630	629	577	645	569	553	614	600	444	387	416	436	447	426
382	404	449	380	423	368	378	395	416	350	386	356	379	393	413	395	406	373	359
395	427	357	412	447	435	385	362	421	395	394	368	449	411	381	406	407	409	365
423	423	357	372	441	407	362	397	356	355	438	378	391	395	422	389	394	448	356
432	365	375	399	356	376	397	418	421	449	398	426	383	384	422	413	424	429	398
439	407	391	374	449	426	447	362	363	409	392	357	374	416	373	359	365	373	388
427	385	417	375	409	351	400	447	378	394	412	381	437	395	386	406	425	422	420
367	382	369	360	380	384	357	419	367	410	356	438	366	410	358	384	441	393	373

Key:

Rounds to 300	Light Blue
Rounds to 400	Dark Blue
Rounds to 500	Gray
Rounds to 600	Black
Rounds to 700	Brown

Name: Date:

Rounding to Hundreds

Knight

550	599	622	564	560	588	633	627	598	442	413	408	602	624	641	582	580	589	561
566	641	571	638	623	559	561	554	438	469	510	543	370	645	603	595	574	636	628
630	593	555	639	600	639	596	612	379	518	409	353	408	573	605	560	625	590	550
570	575	577	617	550	563	635	630	379	460	531	476	384	570	554	635	602	594	575
614	631	554	642	577	639	576	639	436	489	459	532	410	557	644	629	628	585	621
636	602	602	628	362	410	412	384	446	440	358	356	398	412	368	358	609	566	634
561	609	556	444	471	418	493	514	489	498	526	479	457	531	410	460	383	588	578
578	576	402	513	514	356	474	530	453	531	538	483	433	380	359	371	503	399	565
623	394	503	544	396	363	351	540	519	526	450	397	518	506	533	523	373	367	560
421	498	520	368	646	638	359	543	540	453	520	443	450		447	519	369	598	638
406	536	477	389	625	610	440	425	375	353	391	427	463	400		543	383	632	595
354	368	359	395	373	634	399					381	497	468	515	450	417	600	617
570	477	456	644	553	562	438			433			434	452	503	386	632	607	623
622	495	526	587	582	422			409	598	434		435	446	420	373	617	580	609
604	469	475	582	436			350	568	622	647	385			429	604	572	602	588
606	634	508	555	354	375	419	351	626	583	624	604	375	418	385	574	644	579	597
714	726	695	693	698	728	663	737	711	728	729	680	688	702	694	695	696	686	660
728	690	692	402	720	741	655	425	423	689	372	743	735	737	749	670	377	731	668
687	437	401	689	422	664	738	685	745	379	724	686	721	678	394	424	656	436	719
668	739	693	737	680	703	720	688	672	652	682	736	710	653	725	705	744	659	653

Key:

Rounds to 400	Black
Rounds to 500	Gray
Rounds to 600	Blue
Rounds to 700	Green

*Blank squares are red

Name: Date:

Equations

Little Girl

__x3=30	3x__=27	3x__=24	__x3=21	5x__=35	__x2=14	4x__=28	__x4=32	__x7=70	7x__=70
__x2=18	8x__=80	__x10=60	6÷__=3	__÷2=1	10÷__=10	x5=10	4x__=28	9x__=81	80÷__=80
90÷__=9	4x__=40	10x__=80	__x9=27	__x2=2	__x4=4	__x6=24	48÷__=6	8x__=72	__x5=50
90÷__=9	35÷__=5	__x2=2	__x7=7	2x__=4	10÷__=10	7÷__=7	__x8=8	4x__=32	6x__=60
__x4=36	4x__=32	70÷__=10	3x__=3	20÷__=5	16÷__=4	__÷2=1	__x5=35	__x4=32	9x__=90
10x__=100	__x2=2	9x__=54	__x8=40	4x__=8	5x__=5	30÷__=5	25÷__=5	_x5=5	3x__=30
4x__=36	3x__=30	__x8=72	12÷__=2	18÷__=3	42÷__=7	__x7=35	__÷10=1	3x__=27	__x3=30
8x__=80	2x__=18	60÷__=6	10x__=60	10x__=60	__÷6=1	30÷__=5	3x__=27	__x10=100	70÷__=7
72÷__=8	5x__=45	35÷__=7	10x__=50	54÷__=9	9x__=45	54÷__=9	__x9=54	90÷__=9	18÷__=2
20÷__=2	45÷__=5	__x2=18	27÷__=9	12÷__=3	12÷__=4	2x__=6	__x8=80	__x8=80	8x__=72

Key:

1 or 2	Tan
3 or 4	Black
5 or 6	Red
7 or 8	Yellow
9 or 10	Blue

Name: Date:

Equations

Bear

10x__=40	4x__=16	3x__=12	__÷8=1	70÷__=10	32÷__=4	40÷__=5	12÷__=4	__x7=21	__x5=20
20÷__=5	__x3=18	4x__=20	27÷__=9	__x7=28	__x8=24	10x__=40	3x__=15	2x__=12	__x9=36
4x__=16	9x__=54	40÷__=8	__x3=15	60÷__=10	30÷__=6	__x4=24	10x__=50	45÷__=9	__x2=8
32÷__=4	10x__=30	3x__=18	__÷4=1	9x__=54	__x7=35	12÷__=3	9x__=54	8÷__=2	21÷__=3
56÷__=7	30÷__=10	3x__=15	__x6=12	__x8=16	__x6=6	6x__=6	2x__=12	9x__=36	__x2=16
8x__=64	8x__=32	7x__=35	6x__=12	2÷__=2	__x10=10	8÷__=4	__x2=12	18÷__=6	14÷__=2
3x__=21	9x__=27	8x__=48	10÷__=10	30÷__=10	20÷__=5	3x__=6	9x__=54	16÷__=4	__x6=42
__x7=49	64÷__=8	__x5=15	__x9=54	5x__=30	5x__=30	45÷__=9	4x__=12	5x__=40	__x10=80
35÷__=5	6x__=24	10x__=50	__x8=48	50÷__=10	24÷__=4	35÷__=7	20÷__=4	9÷__=3	10x__=80
30÷__=10	7x__=42	5x__=30	6x__=36	50÷__=10	__x5=30	__x8=48	__x3=15	2x__=12	9x__=27

Key:

1 or 2	Tan
3 or 4	Black
5 or 6	Red
7 or 8	Yellow
9 or 10	Blue

Name: Date:

Equations Robot

__x2=14	__x10=80	6÷__=2	49÷__=7	6x__=24	__x9=63	8x__=56	48÷__=6	__x8=64	3x__=24
__x6=48	8x__=56	6x__=48	2x__=6	__x3=21	56÷__=8	__x4=32	8x__=56	__x2=14	10x__=70
5x__=35	__x2=10	__x2=10	18÷__=3	__x3=18	5x__=30	80÷__=10	48÷__=6	2x__=16	8x__=56
__x2=14	__x2=12	__x5=10	20÷__=4	__x9=9	__x5=30	__x4=28	__÷6=1	48÷__=6	__x9=54
35÷__=5	25÷__=5	15÷__=3	__x10=50	6x__=36	54÷__=9	2x__=16	24÷__=4	45÷__=9	10÷__=2
7x__=49	8x__=48	15÷__=3	12÷__=3	2x__=12	40÷__=8	7x__=49	7x =49	10x =60	9x__=72
42÷__=6	__÷7=1	__x9=45	9x__=45	3x__=15	5x__=35	__x8=64	__x6=48	__x10=50	5x__=35
18÷__=3	8x__=48	4x__=24	__x2=10	__x6=30	42÷__=7	__x3=18	__x5=25	3x__=15	__x10=60
48÷__=8	24÷__=4	2x__=8	9÷__=9	__÷3=1	7x__=42	36÷__=6	8x__=56	40÷__=5	__x2=14
50÷__=10	__x8=24	10x__=10	10x__=30	5x__=10	3x__=9	24÷__=4	__x7=49	7x__=56	14÷__=2

Key:

1 or 2	Red
3 or 4	Black
5 or 6	Gray
7 or 8	Blue

Name: Date:

Measurement Choices Beaver

second	minute	day	gallon	quart	cup	cup	second	minute	second
day	second	day	cup	pint		foot	gallon	pint	hour
minute	week	minute	second	gallon	cup	pint	cup		hour
second	hour	minute	minute	quart	quart	quart	day		week
second	minute	day	gallon	quart	pint	pint	day	day	week
meter	mile	day	pint	cup	pint	quart	quart	liter	minute
foot	yard	week	cup	cup	liter	pint	day	week	week
meter	kilometer	meter	pint	liter	liter	quart	gallon	cup	minute
inch	meter	yard	gallon	pint	gallon	gallon	cup	cup	hour
ton	pound	pound	pound	ounce	ton	ounce	ton	pound	pound

Key:

Volume	Brown
Distance	Black
Time	Blue
Weight	Green

*Blank squares are white

Name: Date:

Measurement Choices

Giraffe

week	week	week	minute	hour	ton	hour	week	day	week
hour	day	hour	day	day	ounce	pound		day	hour
minute	minute	week	week	hour	week	liter	pound	ton	pound
hour	second	week	day	week	week	pound	ounce	ton	pound
day	week	minute	second	minute	week	ton	quart	second	minute
hour	second	day	week	minute	second	ounce	ounce	minute	hour
hour	second	day	minute	minute	hour	pint	ton	week	day
pound	minute	minute	minute	week	week	pound	gallon	second	day
inch	ton	pound	quart	ton	ton	pound	ton	meter	mile
yard	cup	ton	ton	ton	cup	pound	quart	foot	mile

Key:

Volume	Brown
Distance	Green
Time	Blue
Weight	Yellow

*Blank squares are black

Name: Date:

Measurement Choices

Surfer

ton	ton	ounce	pound	pound	hour	hour	foot	minute	inch
pound	ounce	ounce	minute	minute	second	foot	minute	yard	second
ounce	ton	week	meter	minute	ounce	hour	minute	week	day
pound	day	inch	second				minute	foot	minute
ton	minute	second	minute	minute		week	day	minute	minute
pound	hour	yard	week	gallon	cup	cup	day	second	second
ounce	inch	hour	week		hour		hour	kilometer	hour
ounce	ounce	second	quart	gallon	cup	quart	foot	week	hour
pound	ounce	pound	second	day	week	yard	day	week	kilometer
ton	ton	ton	ounce	ounce	hour	second	minute	yard	week

Key:

Volume	Brown
Distance	Dark Blue
Time	Blue
Weight	Yellow

*Blank squares are tan

Name: Date:

Identify Shapes Superhero

right	isosceles	equilateral	isosceles	isosceles	equilateral	isosceles	triangle	right	scalene
scalene	isosceles	hexagon	octagon	hexagon	hexagon	octagon	right	triangle	isosceles
triangle	right	rectangle	square	square	square	parallelogram	equilateral	triangle	right
trapezoid	equilateral	rectangle	rectangle	square	trapezoid	rectangle	right	right	equilateral
	right	isosceles	parallelogram	parallelogram	square	scalene	isosceles	right	scalene
									isosceles
triangle	octagon		octagon	hexagon	octagon		octagon		triangle
triangle	hexagon		hexagon	octagon	hexagon		hexagon	parallelogram	scalene
right	hexagon		pentagon	hexagon	pentagon		hexagon	octagon	hexagon
isosceles	hexagon						pentagon	octagon	hexagon

Key:

3 sided shapes	Blue
4 sided shapes	Skin
More than 4 sides	Black

*Blank squares are red

Name: Date:

Identify Shapes Dinosaur

equilateral	equilateral	scalene	equilateral	isosceles	trapezoid		isosceles	triangle	isosceles
scalene	right	triangle	right	parallelogram	trapezoid	trapezoid	trapezoid	right	scalene
isosceles	equilateral	isosceles	triangle	trapezoid	trapezoid	parallelogram	trapezoid	isosceles	scalene
triangle	equilateral	isosceles	isosceles		rhombus	isosceles	isosceles	isosceles	scalene
right	scalene	equilateral	rectangle	trapezoid	rhombus	trapezoid	rectangle	isosceles	triangle
isosceles	isosceles			square	parallelogram	equilateral		isosceles	equilateral
right	rectangle	parallelogram	parallelogram	rhombus	trapezoid	isosceles	isosceles	isosceles	equilateral
rhombus	parallelogram	parallelogram	parallelogram	isosceles	isosceles	isosceles	triangle	triangle	equilateral
triangle	equilateral	scalene	rectangle		triangle	right	isosceles	equilateral	isosceles
pentagon	octagon	hexagon	hexagon	hexagon	octagon	hexagon	hexagon	pentagon	pentagon

Key:

3 sided shapes	Blue
4 sided shapes	Green
More than 4 sides	Brown

*Blank squares are black

Name: Date:

Identify Shapes

Sunflower

isosceles	scalene	isosceles	right	triangle	scalene	isosceles	equilateral	isosceles	triangle
equilateral	triangle	isosceles					scalene	scalene	equilateral
scalene	isosceles		rhombus	square	rhombus	parallelogram		scalene	triangle
scalene	isosceles		rectangle	rectangle	trapezoid	square		triangle	right
isosceles	isosceles		trapezoid	parallelogram	trapezoid	trapezoid		isosceles	isosceles
equilateral	isosceles	right					isosceles	equilateral	isosceles
equilateral	equilateral	right	isosceles	hexagon	octagon	equilateral	isosceles	triangle	scalene
isosceles	scalene	isosceles	triangle	right	hexagon	isosceles	triangle	isosceles	scalene
triangle	triangle	triangle	isosceles	scalene	hexagon	equilateral	right	right	right
hexagon	pentagon	hexagon	hexagon	pentagon	pentagon	hexagon	octagon	octagon	pentagon

Key:

3 sided shapes	Blue
4 sided shapes	Brown
More than 4 sides	Green

*Blank squares are yellow

Name: Date:

Identify Fractions

Heart and Arrow

4/6	4/6	4/6	10/15	6/9	6/9	2/3	2/3	2/3	4/6	4/6	6/9	6/9	4/6	10/15	8/12	8/12	2/3	6/9
10/15	8/12	2/3	2/3	6/9	2/3	10/15	10/15	8/12	10/15	2/3	10/15	8/12	8/12	4/6	4/6			10/15
6/9	4/6	4/6	8/12	10/15	8/12	4/6	4/6	8/12	8/12	8/12	4/6	8/12	10/15	2/3	4/6	9/9		10/15
6/9	4/6	4/6	8/12	8/12	5/15	4/12	2/6	2/3	6/9	10/15	3/9	5/15	5/15	8/12	12/12	8/12	8/12	10/15
4/6	4/6	6/9	6/9	2/6	4/12	4/12	5/15	1/3	8/12	3/9	3/9	4/12	1/3	9/9	4/6	6/9	6/9	6/9
2/3	2/3	2/3	5/15	1/3	2/6	1/3	2/6	4/12	1/3	4/12	5/15	5/15	9/9	2/6	2/6	8/12	10/15	4/6
6/9	4/6	4/12	4/12	2/6	5/15	4/12	5/15	1/3	2/6	2/6	2/6	2/6	2/6	5/15	3/9	4/12	10/15	2/3
10/15	2/3	1/3	4/12	3/9	5/15	3/9	2/6	5/15	2/6	1/3	3/9	2/6	1/3	5/15	1/3	5/15	2/3	8/12
2/3	6/9	1/3	1/3	2/6	2/6	2/6	3/9	1/3	2/6	2/6	4/12	4/12	2/6	2/6	2/6	2/6	4/6	6/9
8/12	8/12	1/3	5/15	1/3	2/6	2/6	4/12	3/9	2/6	3/9	4/12	1/3	4/12	2/6	4/12	2/6	6/9	2/3
4/6	6/9	10/15	2/6	2/6	3/9	1/3	3/9	2/6	1/3	1/3	1/3	3/9	2/6	3/9	5/15	4/6	4/6	6/9
8/12	8/12	6/9	1/3	4/12	4/12	5/15	9/9	2/6	1/3	4/12	1/3	1/3	1/3	1/3	5/15	8/12	8/12	2/3
8/12	2/3	4/6	6/9	2/6	2/6	12/12	5/15	4/12	4/12	1/3	2/6	3/9	4/12	1/3	4/6	10/15	6/9	8/12
10/15	10/15	6/9	4/6	6/9	9/9	4/12	2/6	1/3	2/6	3/9	3/9	5/15	5/15	4/6	4/6	6/9	8/12	6/9
8/12	2/3	8/12	10/15	3/3	6/9	3/9	4/12	5/15	2/6	3/9	2/6	3/9	6/9	2/3	8/12	4/6	2/3	4/6
6/9	6/9		15/15	6/9	2/3	4/6	1/3	2/6	3/9	4/12	2/6	8/12	2/3	10/15	8/12	2/3	10/15	10/15
8/12		15/15		4/6	6/9	4/6	8/12	1/3	5/15	5/15	4/6	6/9	2/3	6/9	2/3	4/6	4/6	2/3
2/3	6/9		6/9	8/12	2/3	4/6	10/15	10/15	3/9	4/6	10/15	8/12	10/15	8/12	8/12	8/12	2/3	8/12
4/6	4/6	4/6	8/12	8/12	8/12	2/3	4/6	8/12	2/3	4/6	4/6	2/3	2/3	4/6	10/15	4/6	4/6	10/15
6/9	10/15	2/3	6/9	8/12	8/12	10/15	6/9	6/9	2/3	4/6	8/12	8/12	2/3	6/9	10/15	4/6	8/12	2/3

Key:

Equal to 1/3	Red
Equal to 2/3	Blue
Equal to 1 whole	Gray

*Blank squares are white

Name: Date:

Identifying Fractions

Spaceship

1/4	2/8	3/12	2/8	3/12	4/16	5/20	1/4	1/4	3/12	3/12	4/16	5/20	5/20	1/4	3/12	3/12	3/12	1/4
1/4	3/12	4/16	5/20	4/16	6/12	1/4	2/8	1/4	10/20	2/8	2/8	1/4	2/4	2/8	3/12	5/20	4/16	1/4
1/4	2/8	2/8	5/20	2/8	3/12	10/20	5/20	1/4	15/20	2/8	4/16	6/12	5/20	2/8	1/4	5/20	3/12	5/20
4/16	2/8	6/12	5/20	3/12	4/16	8/16	10/20	6/8	15/20	9/12	2/4	10/20	1/4	5/20	1/4	6/12	2/8	5/20
2/8	3/12	8/16	4/8	5/20	1/4	2/8	9/12				6/8	4/16	1/4	2/8	2/4	2/4	3/12	2/8
4/16	8/16	4/8	2/4	2/4	6/12	6/8	15/20	3/4	12/16	15/20	3/4	6/8	10/20	4/8	10/20	8/16	6/12	1/4
2/8	10/20	4/8	2/4	3/12	2/4	2/4	12/16	3/4	15/20	12/16	9/12	8/16	10/20	3/12	2/4	10/20	4/8	2/8
3/12				2/8	3/12	1/4	3/12	10/20	2/4	2/4	5/20	5/20	3/12	2/8				4/16
2/8	4/4	4/4	20/20	4/16	2/8	4/16	1/4	3/12	5/20	2/8	4/16	1/4	5/20	4/16	4/4	16/16	8/8	3/12
2/8	8/8	16/16	1/4	4/16	1/4	4/16	2/8	5/20	2/8	1/4	5/20	4/16	1/4	1/4	2/8	8/8	1/4	3/12
3/12	5/20	1/4	4/16	3/12	2/8	1/4	5/20	3/12	3/12	3/12	4/16	4/16	1/4	4/16	3/12	4/16	5/20	2/8
6/12	4/8	8/16	4/8	2/4	4/8	4/16	2/8	1/4	5/20	1/4	2/8	4/16	4/16	5/20	5/20	3/12	5/20	4/16
16/16	16/16	8/16	8/8	4/4	8/16	5/20	10/20	4/8	6/12	8/16	8/16	2/4	8/16	10/20	4/8	6/12	3/12	4/16
8/8	8/8	6/12	16/16	20/20	2/4	3/12	6/12	20/20	4/4	10/20	16/16	8/8	4/8	16/16	8/8	2/4	3/12	3/12
2/4	8/16	10/20	6/12	10/20	6/12	5/20	10/20	4/4	8/8	6/12	4/4	8/8	8/16	12/12	12/12	6/12	3/12	2/8
20/20	16/16	4/8	20/20	16/16	8/16	2/8	6/12	6/12	10/20	6/12	6/12	8/16	6/12	8/16	8/16	4/8	1/4	2/8
12/12	12/12	4/8	20/20	4/4	6/12	3/12	8/16	16/16	20/20	10/20	20/20	8/8	10/20	16/16	20/20	6/12	4/16	3/12
8/16	4/8	6/12	2/4	2/4	6/12	1/4	6/12	4/4	20/20	4/8	4/4	8/8	6/12	4/4	16/16	4/8	1/4	5/20
16/16	20/20	10/20	20/20	16/16	2/4	2/8	10/20	8/16	4/8	8/16	10/20	8/16	6/12	2/4	8/16	8/16	1/4	2/8
8/8	4/4	8/16	4/4	20/20	2/4	4/16	2/4	20/20	12/12	8/16	12/12	12/12	4/8	20/20	12/12	6/12	3/12	1/4

Key:

Equal to 1/4	Blue
Equal to 2/4	Black
Equal to 3/4	Gray
Equal to 1 whole	Yellow

*Blank squares are red

Name: Date:

Identifying Fractions

Mouse

2/10	5/25	5/25	1/5	2/10	5/25	5/25	4/20	5/25	1/5	2/10	3/15	1/5	5/25	2/10	3/15	1/5	5/25	1/5
5/25	3/15	1/5	3/15	5/25	3/15	4/20	3/15	2/10	2/10	1/5	1/5	3/15	1/5	4/20	2/10	5/25	4/20	5/25
3/15	5/25	2/10	2/10	3/15	5/25	4/20	3/15	5/25	4/20	5/25	5/25	5/25	4/20	2/10	2/10	3/15	1/5	5/25
5/25	1/5	4/12	2/6	3/9	5/25	3/15	2/10	3/15	1/5	3/9	3/9	5/15	5/25	5/25	2/10	4/20	1/5	1/5
4/20	3/9	4/8	3/6	3/6	5/15	4/12	4/12	3/9	3/9	1/2	4/8	2/4	4/12	3/15	4/20	3/15	1/5	3/15
5/25	3/9	5/10	5/10	5/15	5/20	1/4	5/20	4/16	5/20	5/15	5/10	2/4	3/9	2/10	3/15	3/15	3/15	4/20
5/25	1/3	1/2	3/9	3/12			2/8			3/12	2/6	2/4	5/15	1/5	3/15	4/20	3/15	4/20
4/20	4/12	4/8	5/15			2/6	5/20	5/15			3/9	3/6	1/3	4/20	1/5	3/9	2/6	5/25
5/25	4/20	1/3	1/3			1/3	4/16	2/6			1/3	5/15	4/20	3/15	5/15	1/4	5/15	1/5
3/15	5/25	5/15	5/20			1/3	3/12	2/6			2/8	3/9	3/15	4/20	1/3	5/20	2/6	1/5
2/10	4/20	3/9	5/20			1/3	3/12	5/15			2/8	1/3	2/10	1/5	2/6	4/16	2/6	2/10
1/5	1/5	1/3	5/20	5/20			1/4			4/16	4/16	4/12	2/10	4/20	2/6	4/16	2/6	1/5
5/25	1/5	2/6	4/16	1/3	4/12	1/3	5/20	2/6	4/12	2/6	3/12	1/3	4/20	2/10	3/9	1/4	1/3	4/20
5/25	5/25	2/10	4/16	5/20	4/16	2/8	2/6	3/12	4/16	2/8	5/20	5/20	2/6	1/3	3/12	5/15	5/25	5/25
4/20	4/20	2/6	2/8	1/3	1/3	5/15	5/20	3/9	5/15	2/6	4/16	5/20	2/8	5/20	5/15	5/25	1/5	3/15
2/10	1/5	3/9	4/16	5/20	2/8	4/16	5/20	3/12	1/4	2/8	4/16	1/4	1/4	3/12	1/3	2/10	5/25	3/15
5/25	3/15	2/6	3/12	2/8	3/9	5/20	3/9	5/20	2/8	2/6	3/12	4/16	2/8	3/12	1/4	2/6	1/5	1/5
5/25	2/10	2/6	2/6	2/6	3/9	1/4	4/12	5/15	5/15	4/12	1/4	3/12	3/12	4/16	4/16	4/12	4/20	2/10
2/10	1/5	4/20	3/15	3/9	1/4	3/12	3/12	3/12	5/20	4/16	1/3	1/4	1/4	4/16	5/20	4/12	5/25	4/20
1/5	2/10	3/15	5/25	3/9	5/15	5/15	4/16	2/8	2/8	3/9	3/9	1/4	5/20	3/12	5/20	4/12	3/15	4/20

Key:

Equal to 1/2	Tan
Equal to 1/3	Black
Equal to 1/4	Brown
Equal to 1/5	Blue

*Blank squares are white

Name: Date:

Identifying Fractions

Handheld

5/10	1/2	4/8	2/6	4/12	3/9	2/6	2/6	1/3	2/6	2/6	2/6	5/15	3/9	5/15	1/2	4/8	3/6	5/10
4/8	4/8	2/4	1/3	2/8	1/4	1/4	3/12	2/8	2/8	2/8	2/8	4/16	5/20	1/3	1/4	3/6	1/2	2/4
5/10	4/8	4/8	2/6	5/20	2/4	2/4	3/6	1/2	5/10	3/6	3/6	3/6	1/4	3/9	3/12	2/4	4/8	2/4
4/8	4/8	5/10	2/6	4/16	3/6	2/4	1/2		3/6	1/2	3/6	1/2	2/8	4/12	1/4	4/8	5/10	3/6
1/2	2/4	3/6	5/15	3/12	5/10	5/10				3/6	2/4	5/10	4/16	1/3	2/8	3/6	2/4	5/10
2/4	4/8	1/2	5/15	5/20	4/8	1/2	3/6	5/10	4/8	1/2	2/4	1/2	3/12	5/15	2/8	1/2	5/10	1/2
1/2	1/2	2/4	5/15	3/12	1/5	3/15	4/8	4/8	2/4	4/8	3/15	1/5	3/12	1/3	1/4	3/6	4/8	3/6
4/8	3/6	4/8	2/6	3/12	1/5	1/5	3/6	3/6	5/10	3/6	2/10	5/25	3/12	4/12	1/4	5/10	3/6	1/2
4/8	4/8	2/4	1/3	5/20	2/8	1/4	5/20	4/16	5/20	2/8	4/16	3/12	4/16	5/15	5/20	4/8	1/2	2/4
3/6	4/8	2/4	2/6	2/6	4/12	3/9	1/3	2/6	4/12	2/6	3/9	3/9	1/3	2/6	2/8	3/6	4/8	3/6
4/8	2/4	2/4	4/12	3/9	1/3	4/12	2/6	3/9	4/12	1/3	1/3	5/15	3/9	1/3	2/8	4/8	1/2	5/10
4/8	5/10	4/8	3/9	4/12	3/15	5/15	5/15	4/12	1/3	2/6	3/9	2/6		1/3	5/20	3/6	3/6	2/4
1/2	5/10	4/8	4/12	3/15	3/15	5/25	4/12	4/12	3/9	1/3		5/15	2/6	4/12	4/16	3/6	1/2	4/8
2/4	4/8	2/4	5/15	1/3	5/25	5/15	3/9	3/9	4/12	3/9	5/15	3/9	3/9	4/12	5/20	3/6	5/10	5/10
3/6	3/6	4/8	2/6	1/3	5/15	1/3	1/3	4/12	2/6	5/15	4/12	2/6	3/9	5/15	5/20	1/2	4/8	4/8
3/6	1/2	1/2	5/15	5/15	3/9	5/15	1/3	4/12	4/12	4/12	4/12	5/15	4/12	2/6	1/4	4/8	5/10	4/8
3/6	5/10	3/6	4/12	4/12	4/12	2/6	2/6	4/12	2/6	2/6	2/6	5/15	1/5	2/6	3/12	3/6	4/8	1/2
5/10	4/8	4/8	5/15	3/9	1/3	3/9	3/9	1/3	4/20	3/9	4/20	2/6	1/3	3/15	1/4	2/4	5/10	3/6
2/4	4/8	2/4	2/6	4/12	3/9	1/3	4/12	5/15	3/9	4/20	1/3	1/5	3/9	5/20	1/2	2/4	2/4	3/6
4/8	4/8	3/6	1/2	5/20	3/12	2/8	1/4	4/16	5/20	3/12	2/8	2/8	4/16	4/8	1/2	5/10	2/4	4/8

Key:

Equal to 1/2	Green
Equal to 1/3	Light Gray
Equal to 1/4	Dark Gray
Equal to 1/5	Black

*Blank squares are red

Key

Name: Date:

Basic Multiplication — City

3x8	3x9	5x6	8x3	6x4	3x9	3x7	7x3	4x7	7x4	10x3	3x10	9x3	3x7	6x5	5x4	5x4	2x7	4x8
3x10	9x3	5x6	5x6	5x5	3x7	6x5	10x3	9x3	3x8	5x5	10x3	8x3	5x6	4x4	3x4	9x2	9x2	5x3
3x7	3x10	6x4	8x3	3x7	5x6	4x6	10x3	7x3	3x7	3x8	7x4	5x5	4x7	6x3	5x8	4x6	6x8	8x5
10x3	3x9	3x10	8x3	9x3	10x3	3x10	9x3	3x10	5x5	6x4	8x3	10x3	4x7	6x3	8x6	3x10	5x10	3x8
3x9	6x5	5x6	3x4	8x2	4x4	6x3	2x10	2x10	2x3	6x5	5x5	10x3	3x9	6x2	8x6	5x8	8x4	10x4
6x5	7x3	3x6	3x2	9x1	2x1	8x1	2x1	2x8	5x3	6x5	3x8	6x4	5x5	2x6	9x4	3x9	4x10	7x3
3x9	3x6	1x10	1x8	1x4	2x1	5x2	2x9	8x0	2x7	3x8	9x3	7x3	3x8	3x5	6x6	5x9	4x9	8x6
3x5	9x2	9x2	3x5	2x10	5x4	4x3	7x0	0x0	8x2	6x5	5x5	9x3	8x3	4x5	7x6	9x3	5x10	6x4
3x4	1x10	3x1	7x1	2x4	1x1	9x2	10x0	1x0	10x2	9x3	5x6	6x5	3x10	10x2	4x9	10x4	7x5	4x10
2x10	9x1	7x4	2x3	5x6	1x4	2x8	2x0	0x4	4x5	6x4	4x6	7x3	9x3	2x10	9x6	3x8	5x9	3x9
2x8	2x1	3x2	1x1	4x2	1x5	2x10	1x0	0x1	4x3	7x2	8x2	5x3	3x5	8x2	7x5	7x7	8x6	5x10
8x2	5x1	8x3	4x2	7x3	9x1	6x3	0x0	10x2	0x1	0x4	0x5	10x0	0x6	5x3	7x2	8x3	6x6	9x3
2x8	1x3	1x9	1x9	1x7	4x1	8x2	6x2	1x0	0x1	0x8	0x7	1x0	1x0	6x2	9x4	5x8	8x4	6x6
2x9	4x1	5x6	1x7	6x4	7x1	4x5	4x5	3x6	4x5	4x4	5x3	3x5	3x4	2x8	4x10	8x3	4x10	3x7
8x2	1x4	2x4	2x1	9x1	1x10	4x4	1x3	1x6	1x6	1x3	3x2	8x1	5x1	6x7	10x5	9x4	10x5	5x10
10x2	4x2	3x9	1x4	6x4	1x1	3x6	1x5	7x3	9x1	4x6	9x1	7x4	1x6	8x2	9x5	4x6	10x4	3x10
2x8	3x1	10x1	8x1	1x1	2x3	3x5	1x10	5x2	5x2	2x5	2x4	4x2	4x1	2x10	5x10	10x5	4x9	7x5
2x6	1x3	9x0	9x0	2x3	2x3	6x3	5x1	10x1	1x2	2x5	0x3	0x3	1x6	2x8	6x8	6x0	0x6	8x5
8x2	10x2	4x5	3x2	5x4	9x2	7x2	2x7	5x4	7x2	2x9	5x4	4x3	5x3	2x10	5x3	10x2	2x10	5x3
2x0	0x9	5x0	0x6	0x6	0x8	0x0	0x6	0x6	0x10	4x0	0x4	10x0	0x9	0x2	0x9	0x2	2x0	8x0

Key:

0	Dark Gray
1-10	Light Gray
11-20	Black
21-30	Light Blue
31-50	Dark Blue

Name: Date:

Basic Multiplication — Cthulu

10x4	6x7	3x4	4x8	6x8	10x2	4x4	3x6	6x2	4x4	2x7	2x9	5x8	4x8	9x5	5x9	10x5	4x10	7x6
10x4	9x5	6x8	8x4	2x10	8x1	4x1	4x2	6x1	9x1	1x2	6x3	7x6	9x5	4x9	8x5	4x9	8x5	4x10
10x5	4x8	7x6	8x4	2x6	8x1	10x3	9x3	4x6	3x10	7x3	3x2	2x6	8x5	4x10	4x9	8x5	10x4	8x6
9x5	7x6	8x5	4x9	5x3	9x1	9x3	6x4	4x7	7x4	9x3	5x2	4x5	9x4	7x7	10x4	5x8	7x5	6x8
9x4	5x7	8x6	7x5	9x2	9x1			9x3			5x1	6x3	7x6	10x5	8x4	5x7	10x5	5x8
6x6	4x8	6x6	7x7	4x3	7x1		2x2	7x3	2x5		3x1	9x2	7x7	4x8	9x5	5x7	9x5	4x9
5x7	7x7	5x9	4x10	9x2	2x3	4x7	10x3	9x3	6x4	7x4	1x6	4x5	5x10	5x7	9x4	4x10	7x6	6x7
4x9	8x6	8x5	4x8	2x10	9x1	5x2	7x3	3x7	7x4	1x1	1x4	4x3	8x4	6x7	10x5	9x5	10x5	4x9
7x5	9x4	5x8	8x4	2x9	7x1	7x1	9x3	9x3	5x6	1x2	1x2	1x7	10x2	10x5	8x5	6x7	6x8	7x7
5x7	5x9	6x6	8x6	8x2	7x1	7x4	7x3	4x7	5x6	10x3	4x1	2x5	10x4	2x10	3x8	7x3	6x7	7x5
7x7	4x8	8x5	2x6	4x4	7x3	3x9	3x7	5x6	6x4	7x3	3x10	3x4	2x10	3x1	5x5	6x4	5x1	5x9
9x5	8x6	5x8	6x3	2x9	3x9	10x2	8x2	3x10	5x4	6x4	6x3	5x6	3x4	7x1	9x2	5x7	5x2	6x6
6x8	9x5	7x5	2x10	2x7	5x5	8x2	6x4	1x3	2x7	6x4	5x4	5x5	2x9	4x2	8x1	5x1	2x5	10x5
10x4	10x4	3x4	4x5	3x4	8x3	10x2	4x6	6x1	3x7	4x4	10x2	3x6	5x4	3x2	2x6	7x2	10x1	8x6
5x10	7x7	5x3	9x2	4x4	9x2	6x2	2x6	1x2	2x9	2x6	2x7	6x3	4x9	6x2	2x0	3x0	2x6	4x8
9x5	6x8	6x5	3x9	10x4	10x2	3x6	2x6	9x1	4x4	2x8	7x2	4x3	5x8	3x4	0x2	4x0	10x2	5x9
8x5	9x4	4x7	7x4	6x6	2x6	9x2	4x4	3x6	2x4	2x7	6x2	10x2	6x8	4x8	4x4	4x3	8x6	8x5
10x4	4x9	5x7	8x6	8x6	2x7	4x5	4x3	6x3	1x2	5x4	6x3	7x7	8x6	9x4	6x6	9x5	9x4	9x4
7x6	7x7	10x4	5x7	8x5	3x6	3x5	6x2	10x2	2x1	9x2	9x2	6x8	4x8	8x5	5x10	4x9	7x5	9x4
4x10	10x5	6x6	8x4	4x9	5x4	6x2	6x2	10x2	5x1	2x9	10x2	4x9	10x5	4x8	6x7	9x4	5x10	5x8

Key:

0	Yellow
1-10	Black
11-20	Brown
21-30	Pink
31-50	Gray

*Blank squares are white

Name: Date:

Basic Multiplication — Apple Tree

3x9	4x7	3x6	8x3	4x6	7x4	3x7	8x3	4x7	4x7	6x5-	1x3	8x5	8x5	4x9	10x4	5x7	8x6	5x9
5x6	3x10	7x3	9x3	6x5	9x3	7x4	7x3	6x4	5x6	7x4	4x2	9x4	6x7	5x7	7x7	9x5	8x4	6x6
5x5	10x3	6x4	8x3	5x6	4x6	7x3	8x3	5x6	1x4	1x1	10x5	4x10	10x5	7x6	8x6	8x6	5x7	8x4
1x2	7x4	7x4	3x7	8x3	10x3	5x5	4x7	1x4	9x5	4x8	5x10	6x8	6x7	6x8	6x7	4x9	8x4	5x10
4x9	2x1	6x1	9x3	7x4	3x10	1x1	1x4	7x5	8x4	1x6	8x5	4x8	2x5	1x6	2x1	2x4	6x8	5x7
7x7	2x3	10x0	1x9	1x7	1x1	6x0	4x1	8x4	10x1	0x7	1x7	2x1	10x3	3x7	6x5	3x10	1x7	7x7
5x10	4x2	0x6	0x10	3x0	0x7	8x0	1x0	4x10	4x2	0x9	0x3	9x1	3x7	6x4	3x10	4x6	3x9	1x1
8x5	1x4	6x0	6x0	0x10	0x0	9x1	6x1	4x2	5x10	5x1	7x0	1x5	3x8	2x5	3x1	1x1	1x8	10x4
8x4	3x2	0x10	2x0	4x0	10x1	2x9	8x2	2x7	4x2	1x7	0x5	1x5	1x6	2x7	3x4	5x4	10x2	2x3
10x5	10x1	2x0	9x0	1x5	7x2		5x3	9x2	6x2	1x10	0x7	5x1	6x3	5x3	2x9			3x1
10x5	3x1	0x2	6x1	9x2		4x5	2x6	7x7	7x6	7x6	1x4	4x7	2x6	7x0	2x8		1x10	10x1
10x5	1x9	0x4	4x1	2x6	4x5	2x7	5x3	2x7	2x10	2x8	3x6	9x2	3x6	2x6			1x5	8x6
4x8	1x5	0x1	7x1	5x3	2x9	2x6	3x5	5x4	6x2	4x3	5x4	4x5	3x4	4x5			5x1	6x8
9x5	5x2	6x0	4x2	8x2	2x8	5x4	3x5	3x6	3x4	5x3	9x2	3x5	2x8	4x3	2x8			2x2
10x5	4x2	4x0	0x6	1x6	8x2	7x2	3x4	9x2	9x2	9x2	9x2	3x6	6x2	2x7	4x5		3x2	8x6
2x5	0x9	0x2	0x8	1x3	6x3	7x2	4x3	7x2	4x3	2x9	2x8	2x10	3x5	3x5	5x3	2x8	6x1	4x8
10x0	0x0	7x0	0x1	0x6	1x2	4x4	2x7	2x7	6x2	6x2	5x4	4x5	6x3	3x5	2x6	7x1	7x6	4x9
0x7	4x1	2x0	8x0	4x2	2x2	5x3	8x2	6x2	6x2	6x2	2x6	4x5	2x9	4x3	1x7	9x3	5x5	7x4
3x2	0x10	1x3	10x1	3x10	10x3	1x5	1x7	9x2	8x2	4x1	1x10	3x4	3x6	1x8	3x10	3x10	4x7	7x4
3x7	5x2	8x3	3x8	6x4	5x5	4x6	3x8	1x2	1x1	3x9	3x8	2x4	8x1	6x4	7x4	3x8	3x9	4x6

Key:

0	Brown
1-10	Black
11-20	Red
21-30	Green
31-50	Blue

*Blank squares are white

Name: Date:

Basic Multiplication — Scooter

8x9	7x9	9x8	8x8	8x10	8x8	9x9	9x10	7x9	8x10	8x8	9x10	10x8	10x10	8x9	10x7	9x10	10x8	8x10
8x10	10x8	9x7	7x9	10x8	8x9	10x9	10x10	8x9	10x9	9x10	9x9	8x9	7x9	8x9	8x8	8x8	9x9	10x10
10x9	10x10	10x10	9x8	7x9	8x8	10x7	2x1	9x1	1x2	1x9	9x8	9x7	8x10	8x10	8x9	10x7	8x10	8x9
8x9	10x9	9x8	9x9	8x8	10x8	2x1	3x1	6x3	6x1	4x5	8x10	9x9	9x9	7x10	9x9	8x10	10x7	8x8
10x10	8x9	7x10	8x10	8x9	7x9	9x10	0x4	10x0	1x0	0x2	9x10	10x10	7x9	7x10	8x10	10x9	10x8	10x9
9x7	9x8	8x8	7x9	10x10	10x9	7x10	0x4	4x0	0x4	8x0	9x9	9x8	8x10	10x10	10x10	8x10	7x10	10x8
8x8	10x7	8x10	5x9	0x5	0x7	6x2	4x4	0x4	6x0	0x7	10x8	9x8	9x10	9x9	9x7	10x10	9x9	9x8
8x9	9x9	10x10	8x6	7x6	4x0	9x0	8x0	6x3	5x3	2x10	9x7	8x10	9x9	9x8	10x9	8x8	8x9	9x9
7x10	7x10	9x9	6x10	10x8	9x10	8x9	10x10	6x2	5x3	2x3	9x8	9x9	8x10	10x9	9x7	8x8	10x10	8x10
7x10	7x9	9x8	6x7	9x9	9x7	9x7	7x9	5x2	2x8	2x7	9x8	8x9	10x9	9x9	10x8	10x9	9x8	7x10
10x10	8x10	7x9	7x6	8x9	10x10	9x10	9x9	10x9			8x10	10x9	8x10	8x9	7x9	10x8	10x8	10x10
8x8	8x10	10x10	10x6	9x8	8x9	9x7	7x9	10x7			10x9	10x10	10x7	0x8	7x10	9x9	9x10	8x10
8x9	8x10	9x9	7x6	8x8	10x7	8x10	8x9	9x8			10x8	7x9	9x7	7x10	10x9	8x9	9x8	7x9
10x10	10x8	10x7	5x9	7x9	10x10	8x8	7x10	7x9			8x9	7x10	7x9	9x8	9x8	10x8	10x10	8x8
10x7	10x8	10x9	8x6	8x10	9x9	10x8	10x7	10x3	7x3	5x6	10x10	10x7	8x8	9x7	7x10	8x9	9x7	10x10
10x10	9x7	10x9	5x9	5x9	9x5	8x6	6x8	6x10	6x9	6x9	7x8	10x6	0x0	0x0	10x10	9x9	10x10	8x9
9x8	9x7	6x6	10x4	3x7	9x7	10x9	8x9	8x10	8x8	8x10	7x9	10x8	10x4	5x6	5x5	8x8	7x10	7x10
9x9	10x8	5x7	5x10	4x7	8x8	10x8	8x10	10x8	10x10	9x9	10x7	8x8	8x3	7x6	4x8	9x8	9x9	7x10
9x9	9x9	3x10	6x5	5x9	10x10	8x8	10x8	7x10	8x8	9x7	10x7	7x10	3x9	9x4	3x10	7x9	10x10	8x9
10x6	6x8	8x6	7x6	8x7	6x10	6x7	7x8	9x6	8x7	6x10	8x6	7x7	8x7	8x6	6x7	9x5	10x5	7x7

Key:

0	Skin
1-20	Red
21-40	Black
41-60	Gray
61-100	Blue

*Blank squares are dark blue

Name: Date:

Basic Multiplication — Flowers

8x8	1x6	2x4	2x8	1x5	1x6	3x1	4x1	3x3	4x2	7x1	3x2	2x3	4x5	1x4	8x2	8x1	5x1	2x9
10x7	7x9	1x6	1x2	5x10	9x6	6x8	6x2	2x10	1x4	4x3	9x2	1x9	4x1	1x5	10x1	5x4	2x2	10x1
7x9	10x9	8x10	7x9	7x9	10x7	8x8	10x9	4x1	4x2	6x3	1x4	9x1	7x7	10x5	6x9	2x8	10x1	10x1
10x9	7x9	8x8	8x10	8x8	10x10	10x10	9x9	8x10	8x6	5x9	9x6	8x6	8x7	6x8	10x5	6x10	9x7	10x8
8x10	10x9	10x7	10x7	10x10	8x9	8x9	9x9	9x8	7x8	6x9	7x8	10x5	10x6	6x8	7x8	9x6	9x10	9x7
10x8	7x10	7x10	9x9	8x8	7x10	9x7	8x10	10x8	6x7	10x6	6x9	6x10	10x5	9x6	9x5	8x6	7x9	10x10
9x10	10x9	9x10	9x10	8x8	9x9	8x8	10x7	8x9	6x10	7x8	9x6	6x7	8x6	8x6	6x9	7x8	9x10	8x10
7x9	9x7	8x8	9x8	9x7	7x9	9x10	8x8	9x7	7x8	6x10	6x9	9x5	7x7	10x4	5x10	7x6	10x9	7x10
9x9	7x10	9x7	9x7	2x0	8x8	8x9	9x8	9x10	7x7	5x9	6x9	5x10	3x10	7x3	5x8	5x10	10x8	7x9
10x7	10x7	9x9	0x10	8x0	0x8	10x7	8x8	10x7	7x6	6x7	9x6	7x6	5x8	8x3	9x3	5x9	7x10	9x8
8x8	10x9	8x9	0x6	0x0	0x10	9x7	7x9	8x10	5x9	6x8	4x9	10x3	3x10	8x0	3x10	9x4	5x5	8x10
9x8	0x9	1x0	2x0	5x7	0x5	0x2	0x0	8x10	9x6	3x7	8x4	6x6	0x6	4x0	0x6	3x7	3x7	9x4
0x8	0x4	9x0	9x3	7x4	4x6	3x0	9x0	5x0	7x7	7x6	8x5	4x7	9x3	0x0	5x6	7x5	10x3	10x9
7x10	0x3	6x0	0x4	6x6	0x4	0x6	3x0	10x9	7x6	5x10	6x10	2x5	4x10	8x5	5x5	7x8	10x9	10x7
10x8	9x7	7x10	9x0	0x0	0x6	2x5	10x10	10x9	9x6	7x8	3x1	10x5	5x6	5x8	7x4	9x6	9x8	7x10
9x7	8x9	10x8	0x10	0x8	10x0	10x10	2x9	10x10	9x5	9x6	4x2	5x9	9x6	7x3	6x9	6x8	9x8	9x7
9x9	9x10	9x10	8x10	0x1	10x7	7x9	10x9	4x4	10x5	6x7	10x1	6x9	6x7	9x5	5x9	9x6	9x10	8x9
10x9	8x10	7x10	10x7	9x10	9x8	10x8	5x9	2x10	9x6	8x6	7x1	6x8	6x10	10x5	10x5	3x6	8x6	7x10
10x7	9x10	7x9	9x7	7x9	8x8	8x7	6x9	2x6	10x6	8x7	2x9	9x6	7x7	8x7	9x6	7x7	8x7	10x7
5x3	3x1	3x1	3x1	3x4	3x1	2x3	5x2	3x2	5x4	1x8	5x2	2x8	2x9	8x2	4x5	9x2	4x4	2x8

Key:

0	Red
1-20	Green
21-40	Yellow
41-60	Brown
61-100	Blue

Name: Date:

Dividing with 1 — Cat

8÷1	10÷1	8÷1	8÷1	8÷1	7÷1	7÷1	10÷1	7÷1	8÷1
10÷1	5÷1	9÷1	7÷1	7÷1	8÷1	10÷1	6÷1	9÷1	8÷1
9÷1	5÷1	5÷1	10÷1	9÷1	10÷1	6÷1	5÷1	10÷1	7÷1
9÷1	5÷1	5÷1	5÷1	5÷1	6÷1	6÷1	6÷1	9÷1	8÷1
9÷1	6÷1		9÷1	5÷1	10÷1		5÷1	10÷1	8÷1
10÷1	6÷1	6÷1	5÷1	6÷1	5÷1	5÷1	5÷1	10÷1	8÷1
10÷1	5÷1	5÷1	6÷1	2÷1	6÷1	6÷1	6÷1	10÷1	7÷1
8÷1	9÷1	5÷1	6÷1	5÷1	5÷1	5÷1	9÷1	7÷1	8÷1
3÷1	4÷1	10÷1	10÷1	9÷1	10÷1	10÷1	3÷1	4÷1	4÷1
4÷1	10÷1	5÷1	5÷1	5÷1	6÷1	5÷1	10÷1	4÷1	3÷1

Key:

1 or 2	Pink
3 or 4	Green
5 or 6	Brown
7 or 8	Blue
9 or 10	Black

*Blank squares are white

Name: Date:

Dividing with 3 — Wizard

30÷3	30÷3	30÷3	27÷3	30÷3	30÷3	30÷3	27÷3	30÷3	30÷3
30÷3	27÷3	27÷3	18÷3	18÷3	30÷3	27÷3	30÷3	27÷3	27÷3
30÷3	27÷3	27÷3	30÷3	18÷3	18÷3	27÷3	30÷3	27÷3	30÷3
27÷3		30÷3	15÷3	18÷3	15÷3	15÷3	27÷3	27÷3	30÷3
30÷3	12÷3	30÷3	27÷3	3÷3	6÷3	30÷3	30÷3	27÷3	30÷3
27÷3	3÷3	18÷3	18÷3			15÷3	18÷3	3÷3	27÷3
30÷3	9÷3	18÷3	15÷3		15÷3	15÷3	18÷3	27÷3	30÷3
27÷3	12÷3	27÷3	15÷3	15÷3	15÷3	18÷3	27÷3	30÷3	30÷3
24÷3	9÷3	24÷3	18÷3	15÷3	15÷3	18÷3	21÷3	24÷3	21÷3
21÷3	21÷3	18÷3	15÷3	15÷3	15÷3	18÷3	15÷3	24÷3	21÷3

Key:

1 or 2	Tan
3 or 4	Brown
5 or 6	Gray
7 or 8	Green
9 or 10	Blue

*Blank squares are white

Name: Date:

Dividing with 5 — Skateboarding

45÷5	50÷5	50÷5	50÷5	50÷5	50÷5	50÷5	50÷5	50÷5	50÷5
45÷5	45÷5	50÷5	50÷5	35÷5	40÷5	50÷5	50÷5	45÷5	45÷5
50÷5	45÷5	50÷5	45÷5	25÷5	25÷5	50÷5	45÷5	45÷5	45÷5
50÷5	45÷5	45÷5	25÷5	20÷5	20÷5	25÷5	50÷5	50÷5	45÷5
50÷5	45÷5	50÷5	50÷5	15÷5	15÷5	50÷5	50÷5	50÷5	50÷5
50÷5	45÷5	50÷5	45÷5	5÷5	5÷5	50÷5	50÷5	50÷5	45÷5
50÷5	45÷5	50÷5	45÷5	5÷5	10÷5	50÷5	50÷5	45÷5	50÷5
50÷5	50÷5	35÷5	35÷5	40÷5	35÷5	40÷5	35÷5	45÷5	50÷5
50÷5	45÷5	50÷5	10÷5	50÷5	50÷5	10÷5	45÷5	50÷5	50÷5
35÷5	35÷5	35÷5	40÷5	40÷5	35÷5	40÷5	35÷5	35÷5	35÷5

Key:

1 or 2	Black
3 or 4	Red
5 or 6	Tan
7 or 8	Brown
9 or 10	Blue

Name: Date:

Dividing with 7 — Campfire

56÷7	56÷7	56÷7	56÷7	63÷7	63÷7	56÷7	63÷7	70÷7	56÷7
49÷7	63÷7	56÷7	49÷7	70÷7	70÷7	56÷7	49÷7	63÷7	56÷7
49÷7	70÷7	63÷7	56÷7	49÷7	56÷7	56÷7	49÷7	56÷7	56÷7
56÷7	63÷7	35÷7	56÷7	42÷7	63÷7	49÷7	56÷7	70÷7	56÷7
63÷7	21÷7	70÷7	56÷7	63÷7	21÷7	70÷7	49÷7	21÷7	63÷7
56÷7	70÷7	28÷7	28÷7	70÷7	21÷7	42÷7	70÷7	21÷7	70÷7
56÷7	21÷7	42÷7	28÷7	21÷7	42÷7	28÷7	28÷7	42÷7	49÷7
56÷7	56÷7	28÷7	35÷7	28÷7	42÷7	28÷7	35÷7	28÷7	49÷7
49÷7	49÷7	21÷7	42÷7	7÷7	7÷7	42÷7	42÷7	28÷7	49÷7
56÷7	7÷7	7÷7	7÷7	21÷7	14÷7	7÷7	7÷7	7÷7	49÷7

Key:

1 or 2	Brown
3 or 4	Red
5 or 6	Yellow
7 or 8	Blue
9 or 10	Gray

Name: Date:

Dividing with 9 — Sailboat

18÷9	9÷9	9÷9	9÷9	18÷9	90÷9	81÷9	54÷9	45÷9	45÷9
9÷9	18÷9	18÷9	18÷9	18÷9	9÷9	90÷9	90÷9	45÷9	54÷9
9÷9	18÷9	9÷9	9÷9	18÷9	18÷9	81÷9	90÷9	81÷9	54÷9
9÷9	9÷9	18÷9	9÷9	9÷9	18÷9	81÷9	90÷9	81÷9	90÷9
81÷9	81÷9	27÷9	81÷9	81÷9	90÷9	90÷9	81÷9	81÷9	90÷9
81÷9	90÷9	27÷9	90÷9	90÷9	81÷9	27÷9	27÷9	81÷9	81÷9
36÷9	27÷9	36÷9	36÷9	27÷9	36÷9	27÷9	81÷9	81÷9	90÷9
27÷9	27÷9	27÷9	27÷9	27÷9	27÷9	63÷9	63÷9	63÷9	63÷9
36÷9	36÷9	27÷9	27÷9	27÷9	72÷9	72÷9	72÷9	63÷9	63÷9
63÷9	63÷9	72÷9	63÷9	63÷9	72÷9	72÷9	72÷9	72÷9	63÷9

Key:

1 or 2	Gray
3 or 4	Brown
5 or 6	Yellow
7 or 8	Dark Blue
9 or 10	Blue

Name: Date:

Basic Division — Baseball

24÷4	30÷5	12÷2	6÷1	12÷2	12÷2	18÷3	12÷2	24÷4	6÷1	24÷4	18÷3	30÷5	12÷2	24÷4	18÷3	18÷3	24÷4	12÷2
12÷2	18÷3	18÷3	18÷3	18÷3	18÷3	18÷3	18÷3	12÷2	12÷2	6÷1	6÷1	6÷1	12÷2	24÷4	30÷5	12÷2	30÷5	24÷4
30÷5	30÷5	18÷3	6÷1	30÷5	30÷5	18÷3	18÷3	24÷4	6÷1	30÷5	18÷3	30÷5	30÷5	6÷1	12÷2	24÷4	30÷5	6÷1
24÷4	18÷3	30÷5	12÷2	30÷5	6÷1	6÷1	6÷1	18÷3	18÷3	18÷3	30÷5	30÷5	30÷5	18÷3	30÷5	24÷4	30÷5	24÷4
6÷1	24÷4	12÷2	12÷2	12÷2	30÷5	30÷5	12÷2	12÷2	30÷5	30÷5	6÷1	6÷1	18÷3	30÷5	12÷2	18÷3	24÷4	6÷1
30÷5	12÷2	24÷4	12÷2	6÷1	6÷1	12÷2			6÷1	30÷5	24÷4	18÷3	6÷1	18÷3	24÷4	30÷5	12÷2	6÷1
12÷2	18÷3	30÷5	30÷5	12÷2	12÷2	30÷5			6÷1	12÷2	24÷4	18÷3	6÷1	9÷3	6÷2	6÷2	15÷5	30÷5
24÷4	18÷3	24÷4	15÷5	6÷2	9÷3	12÷2	6÷1	24÷4	6÷1	30÷5	12÷2	30÷5	12÷4	15÷5	15÷5	15÷5	12÷4	6÷1
24÷4	12÷2	30÷5	12÷4	15÷5	9÷3	15÷5	30÷5	12÷2	12÷2	18÷3	30÷5	18÷3	18÷3	6÷3	2÷1	4÷2	10÷5	24÷4
24÷4	12÷2	12÷2	2÷1	2÷1	2÷1	10÷5	6÷1	24÷4	30÷5	25÷5	15÷3	18÷3	6÷1	4÷2	10÷5	4÷2	4÷2	12÷2
6÷1	24÷4	18÷3	2÷1	2÷1	6÷3	2÷1	24÷4	24÷4	30÷5	5÷1	20÷4	6÷3	15÷5	3÷1	8÷4	6÷3	4÷2	24÷4
16÷4	8÷2	20÷5	2÷1	6÷3	8÷4	16÷4	4÷1				8÷2	8÷2	20÷5	12÷3	9÷3	6÷2	3÷1	16÷4
12÷3	12÷3	2÷1	6÷1	18÷3	30÷5	12÷3	4÷1			16÷4		4÷2	2÷1	8÷4	12÷4	6÷2	9÷3	16÷4
16÷4	4÷1	10÷5	6÷1	12÷2	6÷3	8÷2	16÷4	16÷4	8÷2	20÷5	4÷1		20÷5	16÷4	6÷2	9÷3	12÷4	4÷1
12÷3	16÷4	8÷4	12÷2	30÷5	30÷5	4÷2	2÷1	8÷2	20÷5	4÷1	4÷1	20÷5		8÷2	20÷5	18÷3	30÷5	8÷2
8÷2	8÷2	16÷4	6÷2	12÷4	15÷5	12÷3	4÷1	4÷1	8÷2	20÷5	4÷1	8÷2	4÷1		20÷5	18÷3	6÷1	16÷4
4÷1	16÷4	9÷3	9÷3	8÷2	12÷4	6÷2	4÷1	16÷4	8÷2	4÷1	20÷5	12÷3				6÷1	30÷5	12÷3
16÷4	15÷5	6÷2	20÷5	12÷3	15÷5	12÷4	12÷3	12÷3	4÷1	16÷4	12÷3				4÷4	2÷2	[illegible]	12÷3
	5÷5	5÷5			1÷1	2÷2											20÷5	12÷3
16÷4	4÷1	20÷5	20÷5	8÷2	8÷2	20÷5	16÷4	4÷1	20÷5	16÷4	4÷1	4÷1	4÷1	16÷4	8÷2	4÷1	8÷2	4÷1

Key:

0	Black
1	Skin
2	Red
3	Green
4	Orange
5	Blue

*Blank squares are white

Name: Date:

Basic Division — Lioness

5÷1	20÷4	24÷8	6÷2	9÷3	20÷4	15÷3	50÷10	15÷3	25÷5	15÷3	25÷5	35÷7	40÷8	9÷3	21÷7	3÷1	20÷4	10÷2
40÷8	24÷8	12÷3	4÷1	28÷7	27÷9	30÷6	50÷10	35÷7	15÷3	40÷8	40÷8	40÷8	30÷10	8÷2	16÷4	36÷9	24÷8	10÷2
24÷8	36÷9	4÷1	18÷9	24÷6	4÷1	12÷4	10÷2	15÷3	40÷8	40÷8	20÷4	12÷4	16÷4	8÷2	20÷10	32÷8	8÷2	21÷7
27÷9	28÷7	2÷1	10÷5	14÷7	32÷8	15÷5	5÷1	45÷9	50÷10	5÷1	45÷9	3÷1	20÷5	2÷1	4÷2	2÷1	20÷5	24÷8
15÷5	12÷3	8÷4	20÷10	8÷4	36÷9	27÷9	3÷1	6÷2	27÷9	3÷1	3÷1	21÷7	40÷10	4÷2	10÷5	14÷7	20÷5	18÷6
3÷1	16÷4	20÷5	16÷4	32÷8	12÷3	32÷8	20÷5	24÷6	32÷8	36÷9	20÷5	20÷5	4÷1	4÷1	40÷10	20÷5	40÷10	15÷5
40÷8	12÷4	36÷9	4÷1	20÷5			20÷5	28÷7	36÷9	28÷7	28÷7			40÷10	32÷8	8÷2	9÷3	20÷4
15÷3	12÷4	40÷10	16÷4	16÷4		8÷4	4÷1	12÷3	20÷5	8÷2	8÷2	8÷4		8÷2	8÷2	24÷6	3÷1	20÷4
20÷4	18÷6	28÷7	12÷3	4÷1		10÷5	20÷5	28÷7	36÷9	4÷1	32÷8	4÷2		8÷2	24÷6	40÷10	6÷2	30÷6
50÷10	27÷9	12÷3	4÷1	28÷7	36÷9	12÷3	28÷7	24÷6	24÷6	4÷1	8÷2	8÷2	24÷6	12÷3	12÷3	20÷5	21÷7	15÷3
50÷10	18÷6	10÷10	4÷1	32÷8	20÷5	32÷8	[illegible]	18÷9	10÷5	8÷4	20÷10	24÷6	24÷6	24÷6	4÷1	4÷1	9÷3	20÷4
5÷1	30÷6	15÷5	4÷1	28÷7	40÷10	20÷5	8÷2	4÷2	10÷5	4÷2	32÷8	20÷5	20÷5	28÷7	32÷8	3÷1	25÷5	45÷9
40÷8	50÷10	10÷2	3÷1	40÷10	32÷8	36÷9	4÷4	2÷2	3÷1	7÷7	9÷9	32÷8	40÷10	40÷10	3÷1	10÷2	10÷2	15÷3
25÷5	35÷7	40÷8	35÷7	15÷5	36÷9	24÷8	6÷6	8÷8	21÷7	8÷8	4÷4	18÷6	24÷6	6÷2	15÷3	15÷3	25÷5	15÷3
25÷5	50÷10	24÷8	3÷1	27÷9	8÷2	10÷10	27÷9	9÷3	18÷6	15÷5	9÷3	4÷4	12÷3	27÷9	12÷4	10÷2	30÷6	15÷3
45÷9	6÷2	8÷2	4÷1	9÷3	20÷5	5÷5	1÷1	4÷4	3÷3	9÷9	8÷8	10÷10	12÷3	27÷9	8÷2	27÷9	45÷9	20÷4
9÷3	32÷8	12÷3	20÷5	4÷1	27÷9	18÷6	27÷9	18÷6	3÷1	3÷1	3÷1	18÷6	9÷3	8÷2	12÷3	4÷1	15÷5	20÷4
3÷1	20÷5	12÷3	12÷3	24÷6	10÷10	5÷5	5÷5	6÷6	8÷8	2÷2	2÷2	8÷8	9÷9	12÷3	40÷10	20÷5	24÷6	9÷3
9÷3	36÷9	16÷4	20÷5	4÷4	9÷9	10÷10	9÷9	9÷9	10÷10	3÷3	6÷6	5÷5	10÷10	7÷7	24÷6	28÷7	8÷2	18÷6
30÷10	4÷1	40÷10	40÷10	8÷8	9÷9	8÷8	3÷3	4÷4	1÷1	10÷10	9÷9	10÷10	6÷6	2÷2	4÷1	12÷3	36÷9	15÷5

Key:

1	Tan
2	Black
3	Brown
4	Orange
5	Blue

*Blank squares are white

Name: Date:

Basic Division — Space

20÷4	35÷7	45÷9	20÷10	10÷5	14÷7	6÷3	20÷4	25÷5	20÷4	25÷5	10÷2	15÷3	25÷5	35÷7	25÷5	5÷1	25÷5	15÷3
30÷6	30÷6	24÷8	3÷1	12÷6	12÷6	6÷3	14÷7	15÷3	5÷1	5÷1	30÷6	20÷4	40÷8	10÷2	30÷6	20÷4		35÷7
30÷6	20÷10	3÷1	12÷4	3÷1	20÷10	20÷10	24÷8	18÷6	15÷3	40÷8	15÷3	50÷10	10÷2	50÷10	45÷9	45÷9	25÷5	20÷4
35÷7	20÷10	15÷5	6÷2	15÷5	12÷6	6÷2	4÷2	8÷4	15÷3	15÷3		15÷3	10÷2	10÷2		10÷2	35÷7	10÷2
5÷1	12÷6	2÷1	24÷8	30÷10	6÷3	6÷3	4÷2	10÷5	35÷7	25÷5	20÷4	45÷9	15÷3	50÷10	25÷5	5÷1	25÷5	45÷9
5÷1	10÷5	12÷6	18÷6	8÷4	2÷1	12÷6	6÷2	9÷3	30÷6	30÷6	25÷5	40÷8	50÷10	35÷7	25÷5	20÷4	15÷3	5÷1
25÷5	20÷4	16÷8	14÷7	16÷8	20÷10	18÷6	27÷9	25÷5	20÷4		25÷5	20÷4	30÷6	35÷7	35÷7		40÷8	5÷1
45÷9	15÷3	15÷3	6÷3	18÷9	10÷5	8÷4	35÷7	25÷5	50÷10	5÷1	30÷6	30÷6	15÷3	45÷9	5÷1	50÷10	15÷3	50÷10
	35÷7	45÷9	20÷4	10÷2	50÷10	35÷7	30÷6	5÷1	30÷6	30÷6	25÷5		45÷9	10÷2	50÷10	10÷2	40÷8	45÷9
40÷8	40÷8	50÷10	5÷1	30÷6	45÷9	40÷8		40÷8	25÷5	5÷1	45÷9	15÷3	50÷10	10÷2	25÷5	20÷4	15÷3	45÷9
10÷2	5÷1	40÷8		25÷5	40÷8	35÷7	50÷10	15÷3	40÷8	40÷8	50÷10	40÷8	5÷1	25÷5	40÷8	10÷2	40÷8	5÷1
10÷2	35÷7	35÷7	30÷6	25÷5	40÷8	50÷10	25÷5	15÷3	35÷7		15÷3	45÷9	20÷4	30÷6	25÷5	40÷8	45÷9	35÷7
40÷8	15÷3	20÷4	45÷9	35÷7	10÷2	50÷10	45÷9	15÷3	10÷2	45÷9	50÷10	50÷10	5÷1		30÷6	35÷7	50÷10	30÷6
30÷6	30÷6	30÷6	50÷10		25÷5	50÷10	45÷9	20÷4	20÷4	25÷5	10÷2	35÷7	50÷10	5÷1	30÷6	20÷4	30÷6	35÷7
35÷7	30÷6	5÷1	50÷10	50÷10	25÷5	15÷3	25÷5	40÷8		20÷4	25÷5	35÷7	50÷10	35÷7	25÷5	20÷4	25÷5	10÷2
35÷7	30÷6	5÷1	25÷5	5÷1	20÷4	10÷2	5÷1	35÷7	40÷8	15÷3	10÷2	15÷3	45÷9	40÷8	35÷7	20÷4		35÷7
25÷5		25÷5	30÷6	50÷10	50÷10	15÷3	15÷3	15÷3	15÷3	45÷9	5÷1	5÷1	50÷10	15÷3	50÷10	25÷5	5÷1	15÷3
20÷4	40÷8	5÷1	50÷10	20÷4	45÷9	40÷8	10÷2	10÷2	20÷4	50÷10	50÷10	20÷4	15÷3	45÷9	45÷9	50÷10	30÷6	35÷7
35÷7	25÷5	25÷5	35÷7	12÷3	16÷4	4÷1	40÷10	12÷3	12÷3	12÷3	28÷7	4÷1	4÷1	32÷8	25÷5	10÷2	40÷8	40÷8
40÷8	40÷10	28÷7	32÷8	4÷1	36÷9	28÷7	16÷4	12÷3	4÷1	36÷9	12÷3	36÷9	16÷4	12÷3	4÷1	36÷9	20÷5	10÷2

Key:

1	Blue
2	Green
3	Gray
4	Black

*Blank squares are white

Name: Date:

Basic Division — Cherries

56÷7	56÷8	70÷10	72÷9	48÷6	32÷4	28÷4	21÷3	24÷3	8÷1	72÷9	8÷1	35÷5	63÷9	42÷6	42÷6	42÷6	56÷8	70÷10
35÷5	72÷9	42÷6	21÷3	40÷5	49÷7	70÷10	24÷3	10÷5	1÷1	32÷4	21÷3	70÷10	35÷5	20÷10	7÷7	8÷1	56÷7	64÷8
32÷4	28÷4	7÷1	80÷10	32÷4	21÷3	16÷2	16÷2	56÷8	24÷3	4÷2	49÷7	72÷9	10÷10	60÷10	42÷7	7÷7	70÷10	28÷4
42÷6	48÷6	70÷10	56÷8	80÷10	56÷8	63÷9	7÷1	35÷5	8÷1	10÷10	6÷3	6÷3	60÷10	20÷4	6÷1	36÷6	1÷1	28÷4
56÷8	70÷10	7÷1	80÷10	72÷9	32÷4	72÷9	42÷6	63÷9	10÷10	42÷6	4÷4	7÷7	6÷6	14÷7	3÷3	9÷9	8÷8	7÷1
49÷7	64÷8	14÷2	32÷4	9÷9	5÷5	63÷9	72÷9	3÷3	56÷7	24÷3	32÷4	2÷2	56÷8	24÷3	24÷3	49÷7	72÷9	70÷10
21÷3	56÷8	24÷3	16÷8	27÷9	20÷5	18÷9	3÷3	72÷9	56÷7	21÷3	56÷8	4÷2	14÷2	72÷9	48÷6	49÷7	32÷4	49÷7
35÷5	35÷5	10÷5	28÷7	28÷7	27÷9	12÷3	3÷3	7÷1	56÷8	72÷9	35÷5	70÷10	18÷9	28÷4	14÷2	80÷10	14÷2	21÷3
72÷9	14÷7	20÷5	3÷1	12÷3	4÷1	27÷9	36÷9	10÷10	35÷5	72÷9	80÷10	42÷6	12÷6	24÷3	49÷7	21÷3	28÷4	42÷6
14÷2	16÷8	6÷2	9÷3	28÷7	40÷10		30÷10	2÷1	72÷9	48÷6	48÷6	7÷1	7÷1	18÷9	32÷4	24÷3	49÷7	14÷2
21÷3	63÷9	4÷2	24÷8	28÷7		6÷2	14÷7	21÷3	32÷4	72÷9	49÷7	35÷5	1÷1	14÷7	7÷7	40÷5	63÷9	40÷5
48÷6	56÷8	56÷8	16÷8	24÷6	30÷10	20÷10	56÷8	56÷7	56÷8	32÷4	48÷6	70÷10	14÷7	15÷5	24÷8	2÷2	72÷9	49÷7
35÷5	28÷4	7÷1	8÷1	10÷10	5÷5	32÷4	35÷5	56÷7	80÷10	48÷6	40÷5	4÷2	12÷4	21÷7	30÷10	12÷3	2÷2	42÷6
16÷2	21÷3	24÷3	32÷4	14÷2	72÷9	42÷6	14÷2	49÷7	14÷2	35÷5	14÷7	18÷6	12÷3	40÷10	18÷6	27÷9	20÷5	6÷3
63÷9	56÷8	80÷10	63÷9	42÷6	48÷6	28÷4	64÷8	70÷10	7÷1	21÷3	16÷8	12÷3	28÷7	6÷2	28÷7		3÷1	5÷5
56÷8	56÷8	42÷6	24÷3	21÷3	7÷1	49÷7	24÷3	63÷9	72÷9	21÷3	64÷8	16÷8	30÷10	12÷4		40÷10	10÷10	16÷2
40÷5	63÷9	14÷2	64÷8	80÷10	48÷6	72÷9	8÷1	21÷3	42÷6	21÷3	35÷5	21÷3	3÷3	24÷6	27÷9	20÷10	42÷6	28÷4
7÷1	64÷8	42÷6	40÷5	63÷9	42÷6	70÷10	32÷4	7÷1	49÷7	70÷10	63÷9	64÷8	64÷8	7÷7	9÷9	32÷4	24÷3	48÷6
64÷8	63÷9	7÷1	8÷1	35÷5	21÷3	16÷2	70÷10	8÷1	14÷2	63÷9	56÷8	28÷4	40÷5	70÷10	28÷4	48÷6	7÷1	35÷5
56÷7	56÷7	16÷2	8÷1	16÷2	14÷2	35÷5	35÷5	56÷8	14÷2	42÷6	64÷8	48÷6	49÷7	72÷9	48÷6	8÷1	80÷10	24÷3

Key:

1,2	Black
3,4	Red
5,6	Green
7,8	Blue

*Blank squares are white

Name: Date:

Basic Division — Pig

48÷6	56÷8	5÷5	14÷7	20÷10	64÷8	63÷9	3÷3	2÷1	9÷9	4÷4	21÷3	72÷9	18÷9	18÷9	10÷5	80÷10	80÷10	14÷2
21÷3	7÷1	8÷4	40÷10	20÷5	12÷6	10÷10	28÷7	24÷8	40÷10	27÷9	9÷9	8÷8	4÷1	30÷10	8÷4	49÷7	7÷1	48÷6
40÷5	80÷10	6÷3	28÷7	3÷3	12÷3	12÷4	24÷6	20÷5	16÷4	36÷9	32÷8	4÷1	2÷2	24÷6	18÷9	49÷7	24÷3	8÷1
35÷5	21÷3	49÷7	4÷4	21÷7	27÷9		4÷1	4÷1	3÷1	24÷8		4÷1	8÷2	6÷6	56÷8	80÷10	35÷5	21÷3
49÷7	8÷1	12÷6	6÷2	40÷10	24÷8	12÷6	30÷10	24÷8	40÷10	30÷10	5÷5	32÷8	6÷2	27÷9	4÷2	63÷9	8÷1	7÷1
8÷1	56÷7	12÷6	8÷2	12÷3	32÷8	27÷9	4÷1	3÷1	16÷4	8÷2	28÷7	6÷2	27÷9	16÷4	1÷1	21÷3	14÷2	40÷5
56÷7	21÷3	18÷9	18÷6	18÷6	12÷3	6÷2	30÷10	4÷1	27÷9	40÷10	9÷3	4÷1	40÷10	16÷4	2÷1	32÷4	40÷5	56÷7
14÷2	20÷10	8÷2	4÷1	4÷1	40÷10	30÷10	12÷6	10÷10	18÷9	3÷3	2÷1	3÷3	15÷5	15÷5	12÷3	8÷8	16÷2	24÷3
4÷2	6÷2	9÷3	28÷7	28÷7	24÷6	3÷3	12÷3	8÷2	40÷10	24÷8	9÷3	18÷6	8÷4	32÷8	24÷6	7÷7	28÷4	6÷3
18÷9	9÷3	24÷6	24÷8	28÷7	12÷4	12÷6	12÷4	3÷3	6÷2	28÷7	14÷7	8÷2	1÷1	15÷5	32÷8	18÷9	9÷9	36÷9
4÷4	24÷8	30÷10	3÷1	24÷8	36÷9	2÷1	3÷1	16÷4	9÷3	6÷2	40÷10	32÷8	4÷2	20÷5	9÷3	2÷2	30÷10	10÷10
9÷9	24÷8	24÷8	9÷3	18÷6	18÷6	15÷5	2÷1	2÷1	3÷3	8÷8	20÷10	9÷9	36÷9	15÷5	4÷1	7÷7	18÷9	56÷8
2÷1	18÷6	36÷9	15÷5	12÷4	16÷4	12÷4	28÷7	12÷6	24÷8	40÷10	18÷9	27÷9	16÷4	3÷1	16÷4	10÷10	32÷4	35÷5
9÷9	24÷8	28÷7	40÷10	24÷8	4÷1	9÷3	27÷9	21÷7	6÷3	4÷2	18÷6	12÷4	12÷3	24÷8	15÷5	20÷10	14÷2	40÷5
6÷6	6÷2	12÷4	12÷4	27÷9	20÷5	40÷10	20÷5	16÷4	9÷3	32÷8	3÷1	20÷5	28÷7	20÷5	16÷8	35÷7	24÷4	25÷5
10÷2	4÷2	40÷10	15÷5	20÷5	18÷6	20÷5	32÷8	32÷8	8÷2	27÷9	16÷4	16÷4	6÷2	2÷2	10÷2	24÷4	36÷6	48÷8
25÷5	54÷9	9÷9	36÷9	7÷7	8÷4	3÷1	21÷7	16÷4	8÷2	16÷8	9÷9	40÷10	6÷6	6÷1	30÷5	36÷6	50÷10	45÷9
6÷1	6÷1	20÷4	12÷6	10÷5	6÷6	4÷4	18÷9	1÷1	5÷5	2÷2	8÷4	18÷9	30÷5	45÷9	10÷2	50÷10	5÷1	20÷4
18÷3	25÷5	5÷1	9÷9	9÷9	10÷5	6÷3	18÷3	40÷8	14÷7	7÷7	12÷6	7÷7	25÷5	30÷6	5÷1	54÷9	60÷10	60÷10
60÷10	30÷6	54÷9	30÷5	35÷7	48÷8	6÷1	18÷3	60÷10	45÷9	20÷4	18÷3	12÷2	24÷4	12÷2	48÷8	45÷9	25÷5	48÷8

Key:

1,2	Black
3,4	Pink
5,6	Brown
7,8	Blue

*Blank squares are white

Name: Date:

Addition with Regrouping to Tens — Baby Chick

17 + 15	13 + 32	34 + 14	27 + 13	18 + 28	34 + 14	20 + 29	21 + 18	22 + 20	24 + 26
15 + 11	20 + 18	35 + 15	17 + 39	34 + 19	33 + 27	18 + 39	15 + 33	17 + 22	16 + 22
25 + 13	13 + 31	29 + 42	48 + 19	38 + 21	16 + 46	32 + 34	33 + 19	31 + 17	13 + 36
27 + 22	47 + 13	45 + 23		29 + 39	44 + 30		15 + 43	49 + 19	28 + 12
18 + 23	22 + 31	35 + 24	13 + 46	37 + 28	32 + 43	40 + 14	43 + 11	37 + 29	24 + 12
16 + 14	22 + 46	29 + 31	35 + 36	44 + 34	48 + 49	26 + 32	21 + 43	34 + 31	36 + 11
12 + 14	30 + 37	11 + 45	24 + 39	21 + 49	43 + 26	19 + 50	41 + 23	49 + 25	33 + 14
25 + 21	34 + 30	41 + 23	37 + 15	21 + 46	35 + 27	30 + 21	18 + 47	24 + 34	26 + 19
16 + 24	19 + 16	23 + 36	30 + 24	23 + 28	25 + 48	18 + 34	50 + 12	19 + 15	25 + 16
31 + 12	29 + 16	41 + 43	32 + 47	23 + 22	19 + 13	48 + 39	50 + 47	16 + 24	11 + 28

Key:

From 26 to 50	Blue
From 51 to 75	Yellow
From 76 to 100	Orange

*Blank squares are black

Name: Date:

Addition with Regrouping to Hundreds — Cowboy

252 + 575	250 + 549	422 + 450	220 + 682	302 + 693	177 + 244	220 + 557	364 + 419	469 + 437	109 + 834
566 + 386	720 + 139	180 + 196	649 + 265	191 + 182	245 + 156	214 + 233	637 + 259	218 + 134	423 + 526
299 + 644	326 + 264	233 + 210	131 + 214	163 + 129	166 + 191	225 + 220	214 + 223	154 + 237	548 + 411
422 + 491	610 + 504	429 + 338	403 + 553	346 + 367	324 + 316	364 + 319	472 + 296	243 + 512	348 + 417
711 + 262	715 + 483	381 + 538	193 + 678	200 + 167	307 + 357	187 + 151	330 + 662	598 + 332	490 + 422
504 + 353	821 + 222	229 + 544	246 + 536	330 + 355	562 + 162	421 + 105	408 + 486	483 + 315	606 + 223
431 + 536	528 + 704	245 + 882	190 + 137	455 + 689	614 + 557	707 + 443	148 + 166	474 + 685	896 + 263
422 + 491	711 + 262	504 + 252	177 + 147	159 + 210	262 + 760	239 + 174	238 + 209	473 + 474	557 + 663
578 + 782	348 + 962	384 + 936	141 + 209	154 + 240	976 + 239	176 + 153	228 + 137	635 + 784	433 + 662
379 + 936	938 + 336	549 + 838	183 + 145	159 + 128	180 + 899	163 + 215	232 + 234	750 + 800	456 + 181

Key:

From 251 to 500	Black
From 501 to 750	Skin
From 751 to 1,000	Light Blue
From 1,001 to 1,250	Dark Blue
From 1,251 to 1,500	Green

Name: Date:

Addition with Regrouping to Hundreds — Truck

240 + 529	525 + 252	480 + 518	278 + 483	812 + 135	469 + 323	270 + 641	582 + 369	520 + 374	127 + 782
591 + 176	607 + 151	189 + 694	697 + 181	444 + 332	459 + 353	276 + 591	325 + 507	647 + 318	385 + 441
348 + 434	717 + 150	638 + 132	698 + 145	870 + 298	549 + 605	658 + 252	470 + 352	164 + 804	480 + 510
194 + 753	592 + 301	278 + 483	702 + 224	736 + 365	540 + 409	861 + 294	144 + 744	646 + 272	748 + 250
575 + 480	680 + 570	951 + 108	799 + 327	888 + 154	325 + 815	734 + 374	154 + 915	434 + 629	289 + 528
563 + 514	187 + 207	188 + 275	902 + 200	218 + 073	737 + 204	290 + 209	148 + 255	801 + 395	535 + 425
176 + 775	206 + 243	383 + 107	118 + 815	339 + 537	514 + 316	229 + 240	328 + 172	489 + 308	702 + 205
131 + 519	605 + 145	510 + 160	213 + 298	473 + 207	329 + 240	519 + 122	268 + 300	264 + 435	143 + 634
209 + 476	168 + 482	461 + 148	297 + 354	745 + 372	311 + 381	332 + 280	200 + 430	115 + 475	148 + 815
285 + 264	407 + 198	505 + 211	265 + 375	351 + 268	382 + 294	233 + 366	433 + 267	570 + 316	150 + 614

Key:

From 251 to 500	Black
From 501 to 750	Brown
From 751 to 1,000	Blue
From 1,001 to 1,250	Red

Name: Date:

Subtraction with Regrouping to Tens — Rabbit

48 - 13	47 - 11	74 - 52	44 - 15	60 - 21	91 - 78	71 - 35	81 - 68	58 - 19	63 - 35
73 - 33	93 - 64	49 - 20	96 - 59	84 - 83	69 - 15	49 - 43	65 - 21	52 - 50	46 - 10
75 - 44	48 - 21	76 - 55	41 - 15	69 - 50	87 - 31	65 - 62	85 - 34	49 - 42	95 - 70
27 - 16	60 - 59	34 - 27	22 - 20	26 - 22	84 - 31	63 - 47	61 - 19	63 - 51	42 - 16
72 - 31	99 - 47	86 - 32	37 - 19	72 - 68	84 - 29	84 - 24	63 - 14	91 - 46	25 - 11
97 - 49	78 - 22	74 - 29	96 - 41	56 - 13	94 - 36	64 - 15	63 - 61	68 - 11	22 - 10
95 - 49	92 - 46	80 - 25	72 - 18	62 - 16	77 - 36	91 - 44	81 - 20	90 - 40	86 - 84
80 - 23	55 - 41	36 - 16	74 - 16	98 - 38	89 - 46	60 - 49	96 - 77	75 - 56	94 - 73
00 - 35	55 - 12	79 - 25	79 - 75	75 - 34	92 - 32	69 - 20	80 - 23	26 - 24	73 - 35
80 - 62	65 - 48	75 - 57	48 - 17	74 - 50	81 - 71	86 - 81	53 - 45	62 - 28	96 - 66

Key:

From 1-20	Black
From 21-40	Green
From 41-60	Gray

Name: Date:

Subtraction with Regrouping to Hundreds — Dolphin

985 - 242	802 - 178	873 - 258	884 - 202	929 - 207	930 - 305	915 - 135	827 - 184	953 - 209	891 - 157
921 - 220	566 - 234	901 - 242	921 - 179	993 - 323	941 - 322	940 - 233	987 - 333	976 - 282	840 - 216
829 - 148	897 - 573	910 - 685	832 - 213	876 - 154	945 - 308	917 - 158	876 - 267	766 - 110	884 - 155
445 - 223	629 - 349	833 - 502	695 - 340	907 - 544	830 - 568	967 - 728	821 - 603	838 - 106	856 - 166
575 - 176	370 - 139	997 - 706	727 - 484	559 - 305	926 - 681	656 - 515	890 - 284	638 - 387	779 - 427
844 - 562	967 - 593	706 - 397	476 - 214	473 - 263	525 - 126	636 - 242	646 - 282	911 - 345	554 - 145
370 - 139	560 - 352	312 - 102	937 - 395	778 - 252	506 - 192	860 - 501	994 - 709	937 - 327	827 - 183
926 - 176	841 - 132	835 - 196	955 - 553	898 - 318	857 - 191	909 - 229	928 - 164	927 - 309	898 - 226
933 - 224	902 - 125	805 - 108	665 - 184	981 - 283	778 - 124	958 - 246	950 - 167	839 - 110	919 - 254
914 - 165	872 - 160	734 - 110	978 - 347	864 - 192	875 - 229	909 - 229	811 - 120	807 - 156	924 - 172

Key:

From 1 to 200	Black
From 201 to 400	Light Gray
From 401 to 600	Dark Gray
From 601 to 800	Blue

Name: Date:

Subtraction with Regrouping to Hundreds — Zombie

864 - 204	946 - 294	933 - 275	774 - 128	836 - 146	952 - 265	998 - 310	910 - 134	949 - 203	968 - 336
929 - 147	340 - 182	686 - 550	858 - 172	968 - 346	972 - 268	423 - 318	951 - 912	885 - 211	979 - 196
873 - 260	778 - 505	365 - 144	932 - 141	861 - 165	921 - 255	587 - 171	872 - 276	832 - 230	894 - 176
945 - 320	517 - 286	461 - 148	924 - 549	986 - 255	932 - 479	958 - 547	858 - 267	758 - 121	872 - 141
908 - 225	665 - 346	814 - 582	812 - 112	780 - 134	877 - 223	753 - 201	786 - 186	891 - 218	812 - 210
913 - 143	944 - 863	909 - 817	860 - 655	953 - 276	877 - 454			721 - 268	941 - 298
887 - 134	459 - 441	988 - 802	899 - 279	970 - 216	912 - 266			925 - 196	902 - 259
922 - 281			953 - 197	863 - 138	970 - 233	803 - 616	201 - 175	787 - 171	763 - 113
776 - 157			943 - 248	997 - 335	793 - 150	352 - 159	581 - 478	869 - 190	908 - 295
780 - 549	799 - 416	877 - 546	954 - 594	973 - 755	492 - 231	859 - 478	598 - 371	956 - 695	804 - 589

Key:

From 1 to 200	Black
From 201 to 400	Green
From 401 to 600	Skin
From 601 to 800	Blue

*Blank squares are purple

Name: Date:

Place Value Multiplication — Fun Alligator

4x80	4x80	8x50	4x80	4x90	9x40	6x60	4x90	8x50	7x50	5x70	4x80	8x40	5x80	4x90	8x40	4x90	7x50	8x40
4x90	4x80	4x80	4x90	8x50	5x70	4x80	8x40	4x90	5x70	5x70	7x50	5x80	6x60	8x40	6x60	4x80	4x80	6x60
8x50	5x80	5x80	4x90	4x90	6x50	4x60	7x40	5x70	4x60	3x80	5x80	8x40	8x50	5x70	5x70	5x80	4x80	5x70
5x70	4x90	8x40	6x60	4x70	9x30			5x60			5x50	7x50	8x50	4x80	4x90	4x90	5x70	5x70
8x40	5x80	7x50	3x70	6x40		1x60	1x30	3x70	2x10		7x30	9x40	4x90	7x50	6x60	7x50	4x90	5x80
8x40	5x70	4x90	9x40	3x80		3x30	2x20	4x70	3x30		6x40	8x50	7x50	3x90	3x70	9x30	6x60	4x90
9x40	7x50	6x60	5x50	5x50	4x70			8x30		6x40	9x30	7x30	3x70	1x50	6x50	1x90	9x30	8x40
9x40	4x90	4x90	4x90	7x40	8x30	5x60	3x90	8x30	7x40	3x80	7x30	9x30	9x30	8x30	6x50	7x40	6x40	9x40
5x80	5x70	8x40	9x30	4x70	3x40	4x50	3x50	9x20		2x90		6x20		4x50		6x60	6x60	6x60
5x70	9x40	5x70	5x80	7x40	9x30	2x90	9x20	4x40	9x20	5x30	8x20	8x40	5x70	5x70	9x40	8x50	9x40	4x90
8x40	8x40	8x40	3x90	8x30	5x50	3x40	5x40	3x60	5x30	7x20	6x60	5x70	4x90	6x60	8x40	9x40	5x80	4x80
5x80	8x50	7x50	4x90	6x40	6x40	2x70	8x20	4x50	3x40	6x60	4x70	9x40	5x80	4x80	5x70	5x80	6x60	8x50
5x80	8x50	6x60	5x60	6x50	7x30	6x40	4x60	5x60	6x50	3x70	8x30	3x70	8x50	8x40	8x40	8x50	6x60	5x70
4x80	7x50	6x60	8x40	6x50	5x50	8x30	4x60	4x60	3x90	4x70	3x80	3x90	7x50	8x40	8x40	8x50	7x50	4x90
5x70	4x80	4x80	3x90	8x30	5x60	3x40	2x70	8x20	5x40	6x60	5x80	5x70	4x90	4x90	4x90	4x90	4x90	5x70
8x50	8x50	6x60	5x70	3x90	9x30	5x30	7x20	4x40	7x20	5x30	4x90	8x50	4x90	5x70	8x50	5x70	4x80	5x80
8x50	4x90	5x80	3x80	5x50	6x50	9x30	3x60	4x40	3x60	3x50	5x70	4x90	5x80	6x60	7x50	9x40	9x40	4x80
3x90	5x70	7x40	5x70	6x50	7x40	6x50	5x40	2x80	3x50	8x20	8x30	5x70	5x70	4x90	4x80	9x40	7x50	4x90
5x50	4x70	9x30	3x90	3x90	7x40	5x50	8x30	5x50	2x90	3x70	4x60	8x40	8x40	9x40	8x40	5x80	8x50	6x60
3x70	7x30	6x40	5x60	4x60	5x60	3x70	8x30	5x60	6x60	4x60	6x50	7x50	9x40	4x80	6x60	8x40	8x50	9x40

Key:

1-100	Black
101-200	Light Green
201-300	Dark Green
301-400	Blue

*Blank squares are white

Name: Date:

Place Value Multiplication — Basketball

				7x80	6x90	8x60	8x80	9x70	5x90	8x90	6x90	6x70	7x70	7x80	8x60	8x90	6x90	8x80
					7x60	8x90	9x90	6x70	8x90	8x90	6x70	6x70	6x90	6x90	9x60	8x90	8x90	7x90
					9x50	9x70	7x70	8x90	8x70	6x70	6x80	8x90	6x90	7x80	7x90	7x60	5x90	9x60
		7x40	3x70	7x40	8x60	8x70	6x80	5x90	6x90	8x70	9x80	6x90	8x70	8x60	7x60	5x90	8x90	8x70
	3x80				8x30	5x90	6x70	7x70	7x70	9x50	8x60	5x90	6x80	8x60	9x60	6x70	7x60	8x70
		7x30	7x30	6x40	6x70	8x70	7x80	5x90	9x80	8x60	9x80	6x80	5x20	3x20	1x60	8x10	1x10	7x80
7x60	7x10	6x80	8x90	8x90	6x80	8x60	7x80	9x60	7x70	6x70	9x80	9x50	7x60	2x10	3x10	1x50	1x80	6x70
6x80	7x10	7x80	9x50	6x90	8x90	9x50	7x70	6x80	7x60	8x70	6x90	8x80	6x60	4x90	4x90	4x20	1x50	4x10
9x80	9x10	8x90	5x90	9x70	8x90	6x90	8x60	7x90	5x60	9x30	6x90	9x70	8x40	5x80	8x40	1x60	1x90	9x10
6x70	8x10	9x70	9x80	9x50	7x60	6x90	8x60	3x90	3x80	3x80	4x70	8x50	6x40	7x50	4x90	9x40	1x30	7x10
9x60	3x20	6x90	7x60	9x90	7x80	8x80	9x70	9x30	9x30	7x30	5x60	7x70	8x80	5x60	4x60	3x70	1x50	1x90
8x80	5x20	7x70	5x90	6x80	7x60	6x70	8x70	8x80	8x30	9x30	7x50	7x50	4x80	7x40	9x30	7x30	4x10	7x60
6x70	1x50	7x60	7x80	6x70	9x60	9x80	8x70	7x60	8x60	9x70	8x90	8x90	9x90	3x70	6x40	6x40	5x90	9x60
9x60	4x20	7x70	6x80	6x70	8x60	7x70	8x90	7x80	6x80	9x60	9x50	8x80	6x70	7x60	1x90	1x90	8x90	9x70
2x70	1x50	7x20	7x20	5x40	2x80	8x20	4x50	6x50	4x40	2x70	2x90	8x20	2x70	5x30	3x10	1x50	2x90	5x30
4x50	1x60	3x50													4x80	7x50		
3x40	7x10		2x60	4x30	3x50	3x50	9x20	8x20	2x90	6x30	5x40	2x90	3x50	3x20	1x80	3x30	4x40	4x40
6x30		3x40	6x30	8x20	6x20	8x20	4x40	2x70	8x20	8x20	6x20	3x40	5x30	4x50	3x50	9x20	5x30	2x60
	4x30	4x40	4x50	2x90	4x50	4x30	3x60	7x20	5x40	4x50	4x50	3x40	9x20	8x20	7x20	4x30	4x30	6x20
2x60	4x30	9x20	3x60	2x70	3x50	8x20	3x40	5x40	4x30	6x30	5x30	3x60	5x40	2x70	3x50	3x60	3x40	5x30

Key:

1-100	Black
101-200	Brown
201-300	Orange
301-400	Skin
401-1,000	Blue

*Blank squares are white

Name: Date:

Place Value Multiplication — Castle

9x40	6x60	4x80	5x70	8x40	6x60	4x80	5x70	4x80	6x60	5x70	7x50	4x90	5x70	9x40	8x40	8x40	6x60	8x50
9x40	1x20	4x90	2x20	9x40	5x20	8x40	5x70	7x50	7x50	8x40	6x60	6x60	9x10	7x50	1x40	5x80	2x20	9x40
1x10	7x40	4x10	6x50	1x70	3x90	9x10	8x50	6x60	6x60	4x90	5x70	3x20	3x80	1x70	6x40	5x10	3x70	6x10
1x30	9x30	7x30	3x70	5x50	3x80	1x70	4x90	8x50	4x80	9x40	4x90	3x20	7x30	7x30	3x90	3x80	3x90	2x50
6x60	1x30	2x20	3x20	7x10	5x10	4x90	5x80	9x40	5x80	5x70	6x60	8x40	1x90	2x20	1x90	1x90	1x20	7x50
8x50	4x80	1x80	8x30	3x30	6x60	5x70	7x50	9x40	5x70	8x50	4x90	8x40	4x90	1x50	7x40	1x50	5x80	5x80
6x60	8x50	1x20	8x30	3x20	2x40	5x70	1x90	5x70	4x20	5x70	1x50	4x90	1x10	2x20	7x40	2x10	4x90	8x40
4x90	5x80	9x10	6x40	2x50	3x80	3x20	3x80	1x50	3x90	1x60	7x40	5x10	5x50	3x20	5x50	1x90	4x90	4x80
7x50	4x90	3x20	4x70	1x50	9x30	9x30	3x90	5x60	8x30	4x70	3x70	9x30	7x30	2x50	7x30	3x20	8x50	4x90
4x80	4x90	2x30	3x70	9x10	6x40	8x30	4x60	3x70	9x30	4x60	4x60	3x70	3x90	3x10	3x70	1x10	5x80	6x60
8x50	5x70	1x60	8x30	2x20	9x30	6x50	3x90	6x40	5x50	3x90	4x70	7x40	8x90	8x80	6x50	5x10	7x50	9x40
5x80	9x60	7x90	8x30	1x90	6x50	5x50	2x40	9x10	8x30	3x90	9x30	9x90	9x60	8x90	5x90	2x20	7x50	6x60
7x70	9x70	9x50	6x70	8x10	7x40	2x50	3x10	2x30	1x60	4x70	7x30	7x70	5x90	8x70	8x90	4x10	9x40	8x40
6x70	9x80	9x60	6x90	4x20	5x50	6x10	2x30	5x10	1x50	4x70	9x30	7x70	9x50	7x90	9x50	3x10	6x60	5x80
6x80	7x60	8x60	7x80	1x80	7x30	4x10	1x70	2x40	3x10	3x90	3x70	9x90	6x90	6x80	6x90	3x30	5x80	5x80
7x50	6x30	6x30	3x70	1x40	9x30	1x20	1x30	5x10	4x20	5x50	3x70	3x70	3x40	2x90	5x50	1x70	4x80	8x40
5x10	3x50	3x40	4x10	5x20	4x20	4x10	1x10	8x10	1x40	1x10	1x30	1x50	3x60	3x40	1x70	1x80	9x10	1x90
7x70	7x90	9x60	7x80	7x60	7x70	7x70	8x60	8x70	8x80	5x90	7x80	[illegible]	3x60	3x60	[illegible]	7x70	7x80	7x70
9x80	7x90	6x70	5x90	8x80	9x60	7x60	7x90	9x50	9x70	6x80	8x70	9x50	4x50	6x30	9x80	9x90	9x50	9x80
5x90	8x70	8x70	9x50	9x60	9x80	8x60	9x50	6x80	7x80	9x50	8x70	7x70	7x90	6x80	6x80	9x60	6x70	7x90

Key:

1-100	Black
101-200	Brown
201-300	Gray
301-400	Blue
401-1,000	Green

Name: Date:

Rounding to Tens — Fishbowl

29	29	27	27	31	28	34	31	28	29	32	32	30	31	34	34	27	32	27
28	29	33	25	25	33	28	27	29	32	30	33	34	30	25	29	26	27	34
32	26	30	34	29	30	31	34	29	33	34	26	30	27	34	32	31	32	28
30	26	28	6	13	6	6	10	7	6	5	13	10	10	12	14	27	32	31
28	26	33	14	14	7	12	5	12	8	12	8	9	9	14	10	32	30	28
32	32	27	31	5	10	14	7	6	6	9	12	9	8	7	26	27	31	29
32	30	28	7	5	8	14	9	13	9	8	5	13	11	5	9	26	28	28
33	34	14	6	13	12	10	14	10	13	9	5	9	7	10	11	9	25	30
32	28	18	21	15	17	15	22	17	15	20	22	22	22	16	19	20	33	31
31	29	21		16				20	16	21	23	19	16	21	15	19	34	28
28	32	22					41		16	18	22	16	15	17	18	24	32	26
30	29	18							17	23	23	24	19	20	24	16	29	33
26	27	22		24				17	15	41		24		17	17	17	31	26
27	26	22	23	15	16	22	17	23						15	20	24	28	29
32	28	32	19	15	15	24	18	18	22			17		17	16	28	29	26
30	27	28	27	23	19	21	15	22	18	20	24	18	19	21	31	30	30	31
38	37	41	38	36	35	39	41	40	35	43	36	44	38	42	42	38	40	44
39	36	37	43	41	35	37	36	38	43	37	40	35	41	41	43	39	37	44
38	38	36	35	37	36	39	39	35	40	38	35	35	44	42	35	40	40	37
38	35	37	36	35	44	44	41	37	36	44	39	39	42	40	39	40	41	43

Key:

Rounds to 10	Light Blue
Rounds to 20	Dark Blue
Rounds to 30	Gray
Rounds to 40	Black

*Blank squares are orange

Name: Date:

Rounding to Hundreds — Hockey Stick

302	335	259	323	287	320	302	264	280	305	280	335	343	293	310	269	265	314	303
321	273	318	299	349	272	321	313	268	347	282	331	286	299	322	274	335	280	715
281	341	336	278	320	315	285	335	306	271	268	336	317	348	256	312	281	720	683
272	314	272	287	294	312	320	272	323	299	342	293	301	330	299	308	695	663	654
283	330	666	695	736	667	653	680	728	747	690	689	717	747	729	688	703	714	721
279	657	693	651	650	740	733	719	668	653	709	700	658	658	725	673	663	706	652
278	650	692	650	690	722	681	730	737	675	650	722	723	685	747	667	721	717	260
290	726	731	532	548	479	514	479	525	497	481	470	463	663	697	726	742	713	256
348	738	471	530	499	547	513	497	465	541	501	479	485	510	724	650	735	300	306
331	688	465	485	519	490	486	491	469	513	514	524	515	512	674	650	311	318	327
431	421	580	488	488	496	547	484	508	525	454	513	505	560	442	393	422	354	430
419	365	572	613	599	593	640	577	608	573	611	568	577	607	423	387	409	394	408
432	432	404	619	641	630	629	577	645	569	553	614	600	444	387	416	436	447	426
382	404	449	380	423	368	378	395	416	350	386	356	379	393	413	395	406	373	359
395	427	357	412	447	435	385	362	421	395	394	368	449	411	381	406	407	409	365
423	423	357	372	441	407	362	397	356	355	438	378	391	395	422	389	394	448	356
432	365	375	399	356	376	397	418	421	449	398	426	383	384	422	413	424	429	398
439	407	391	374	449	426	447	362	363	409	392	357	374	416	373	359	365	373	388
427	385	417	375	409	351	400	447	378	394	412	381	437	395	386	406	425	422	420
367	382	369	360	380	384	357	419	367	410	356	438	366	410	358	384	441	393	373

Key:

Rounds to 300	Light Blue
Rounds to 400	Dark Blue
Rounds to 500	Gray
Rounds to 600	Black
Rounds to 700	Brown

Name: Date:

Rounding to Hundreds — Knight

550	599	622	564	560	588	633	627	598	442	413	406	602	624	641	582	580	589	561
566	641	571	638	623	559	561	554	438	469	510	543	370	645	603	595	574	636	628
630	593	555	639	600	639	596	612	379	518	409	353	408	573	605	560	625	590	550
570	575	577	617	550	563	635	630	379	460	531	476	384	570	554	635	602	594	575
614	631	554	642	577	639	576	639	436	489	459	532	410	557	644	629	628	585	621
636	602	602	628	362	410	412	344	446	440	358	356	398	412	368	358	609	566	634
561	609	556	444	471	438	493	514	489	498	526	479	457	531	410	460	383	588	578
578	576	402	513	514	356	474	530	453	531	538	483	433	380	359	371	503	399	565
623	394	503	544	396	363	351	540	519	526	450	397	518	506	533	523	373	357	560
421	498	520	368	646	638	359	543	540	453	520	443	450		447	519	369	598	638
406	536	477	389	625	610	440	425	375	353	391	427	463	400		543	383	632	595
354	368	359	395	373	634	399					381	497	468	515	450	417	600	617
570	477	456	644	553	562	438			433			434	452	503	386	632	607	623
622	495	526	587	582	422			409	598	434		435	446	420	373	617	580	609
604	469	475	582	436			350	568	622	647	385			429	604	572	602	588
606	634	508	555	354	375	419	351	626	583	624	604	375	418	385	574	644	579	597
714	726	695	693	698	728	663	737	711	728	729	680	688	702	694	695	696	686	660
728	690	692	402	720	741	655	425	423	689	372	743	735	737	749	670	377	731	668
687	437	401	689	422	664	738	685	745	379	724	686	721	678	394	424	656	436	719
668	739	693	737	680	703	720	688	672	652	682	736	710	653	725	705	744	659	653

Key:

Rounds to 400	Black
Rounds to 500	Gray
Rounds to 600	Blue
Rounds to 700	Green

*Blank squares are red

Name: Date:

Equations — Little Girl

__x3=30	3x__=27	3x__=24	__x3=21	5x__=35	__x2=14	4x__=28	__x4=32	__x7=70	7x__=70
__x2=18	8x__=80	__x10=60	6÷__=3	__÷2=1	10÷__=10	__x5=10	4x__=28	9x__=81	80÷__=80
90÷__=9	4x__=40	10x__=80	__x9=27	__x2=2	__x4=4	__x6=24	48÷__=6	8x__=72	__x5=50
90÷__=9	35÷__=5	__x2=2	__x7=7	2x__=4	10÷__=10	7÷__=7	__x8=8	4x__=32	6x__=60
__x4=36	4x__=32	70÷__=10	3x__=3	20÷__=5	16÷__=4	__÷2=1	__x5=35	__x4=32	9x__=90
10x__=100	__x2=2	9x__=54	__x8=40	4x__=8	5x__=5	30÷__=5	25÷__=5	__x5=5	3x__=30
4x__=36	3x__=30	__x8=72	12÷__=2	18÷__=3	42÷__=7	__x7=35	__÷10=1	3x__=27	__x3=30
8x__=80	2x__=18	60÷__=6	10x__=60	10x__=60	__÷6=1	30÷__=5	3x__=27	__x10=100	70÷__=7
72÷__=8	5x__=45	35÷__=7	10x__=50	54÷__=9	9x__=45	54÷__=9	__x9=54	90÷__=9	18÷__=2
20÷__=2	45÷__=5	__x2=18	27÷__=9	12÷__=3	12÷__=4	2x__=6	__x8=80	__x8=80	8x__=72

Key:

1 or 2	Tan
3 or 4	Black
5 or 6	Red
7 or 8	Yellow
9 or 10	Blue

Name: Date:

Equations — Bear

10x__=40	4x__=16	3x__=12	__÷8=1	70÷__=10	32÷__=4	40÷__=5	12÷__=4	__x7=21	__x5=20
20÷__=5	__x3=18	4x__=20	27÷__=9	__x7=28	__x8=24	10x__=40	3x__=15	2x__=12	__x9=36
4x__=16	9x__=54	40÷__=8	__x3=15	60÷__=10	30÷__=6	__x4=24	10x__=50	45÷__=9	__x2=8
32÷__=4	10x__=30	3x__=18	__÷4=1	9x__=54	__x7=35	12÷__=3	9x__=54	8÷__=2	21÷__=3
56÷__=7	30÷__=10	3x__=15	__x6=12	__x8=16	__x6=6	6x__=6	2x__=12	9x__=36	__x2=16
8x__=64	8x__=32	7x__=35	6x__=12	2÷__=2	__x10=10	8÷__=4	__x2=12	18÷__=6	14÷__=2
3x__=21	9x__=27	8x__=48	10÷__=10	30÷__=10	20÷__=5	3x__=6	9x__=54	16÷__=4	__x6=42
__x7=49	64÷__=8	__x5=15	__x9=54	5x__=30	5x__=30	45÷__=9	4x__=12	5x__=40	__x10=80
35÷__=5	6x__=24	10x__=50	__x8=48	50÷__=10	24÷__=4	35÷__=7	20÷__=4	9÷__=3	10x__=80
30÷__=10	7x__=42	5x__=30	6x__=36	50÷__=10	__x5=30	__x8=48	__x3=15	2x__=12	9x__=27

Key:

1 or 2	Tan
3 or 4	Black
5 or 6	Red
7 or 8	Yellow
9 or 10	Blue

Name: Date:

Equations — Robot

__x2=14	__x10=80	6÷__=2	49÷__=7	6x__=24	__x9=63	8x__=56	48÷__=6	__x8=64	3x__=24
__x6=48	8x__=56	6x__=48	2x__=6	__x3=21	56÷__=8	__x4=32	8x__=56	__x2=14	10x__=70
5x__=35	__x2=10	__x2=10	18÷__=3	__x3=18	5x__=30	80÷__=10	48÷__=6	2x__=16	8x__=56
__x2=14	__x2=12	__x5=10	20÷__=4	__x9=9	__x5=30	__x4=28	__÷6=1	48÷__=6	__x9=54
35÷__=5	25÷__=5	15÷__=3	__x10=50	6x__=36	54÷__=9	2x__=16	24÷__=4	45÷__=9	10÷__=2
7x__=49	8x__=48	15÷__=3	12÷__=3	2x__=12	40÷__=8	7x__=49	7x__=49	10x__=60	9x__=72
42÷__=6	__÷7=1	__x9=45	9x__=45	3x__=15	5x__=35	__x8=64	__x6=48	__x10=50	5x__=35
18÷__=3	8x__=48	4x__=24	__x2=10	__x6=30	42÷__=7	__x3=18	__x5=25	3x__=15	__x10=60
48÷__=8	24÷__=4	2x__=8	9÷__=9	__÷3=1	7x__=42	36÷__=6	8x__=56	40÷__=5	__x2=14
50÷__=10	__x8=24	10x__=10	10x__=30	5x__=10	3x__=9	24÷__=4	__x7=49	7x__=56	14÷__=2

Key:

1 or 2	Red
3 or 4	Black
5 or 6	Gray
7 or 8	Blue

Name: Date:

Measurement Choices — Beaver

second	minute	day	gallon	quart	cup	cup	second	minute	second
day	second	day	cup	pint		foot	gallon	pint	hour
minute	week	minute	second	gallon	cup	pint	cup		hour
second	hour	minute	minute	quart	quart	quart	day		week
second	minute	day	gallon	quart	pint	pint	day	day	week
meter	mile	day	pint	cup	pint	quart	quart	liter	minute
foot	yard	week	cup	cup	liter	pint	day	week	week
meter	kilometer	meter	pint	liter	liter	quart	gallon	cup	minute
inch	meter	yard	gallon	pint	gallon	gallon	cup	cup	hour
ton	pound	pound	pound	ounce	ton	ounce	ton	pound	pound

Key:

Volume	Brown
Distance	Black
Time	Blue
Weight	Green

*Blank squares are white

Name: Date:

Measurement Choices — Giraffe

week	week	week	minute	hour	ton	hour	week	day	week
hour	day	hour	day	day	ounce	pound		day	hour
minute	minute	week	week	hour	week	liter	pound	ton	pound
hour	second	week	day	week	week	pound	ounce	ton	pound
day	week	minute	second	minute	week	ton	quart	second	minute
hour	second	day	week	minute	second	ounce	ounce	minute	hour
hour	second	day	minute	minute	hour	pint	ton	week	day
pound	minute	minute	minute	week	week	pound	gallon	second	day
inch	ton	pound	quart	ton	ton	pound	ton	meter	mile
yard	cup	ton	ton	ton	[illegible]	pound	quart	[illegible]	mile

Key:

Volume	Brown
Distance	Green
Time	Blue
Weight	Yellow

*Blank squares are black

Name: Date:

Measurement Choices — Surfer

ton	ton	ounce	pound	pound	hour	hour	foot	minute	inch
pound	ounce	ounce	minute	minute	second	foot	minute	yard	second
ounce	ton	week	meter	minute	ounce	hour	minute	week	day
pound	day	inch	second				minute	foot	minute
ton	minute	second	minute	minute		week	day	minute	minute
pound	hour	yard	week	gallon	cup	cup	day	second	second
ounce	inch	hour	week		hour		hour	kilometer	hour
ounce	ounce	second	quart	gallon	cup	quart	foot	week	hour
pound	ounce	pound	second	day	[illegible]	yard	day	week	kilometer
ton	ton	ton	ounce	ounce	hour	second	minute	yard	week

Key:

Volume	Brown
Distance	Dark Blue
Time	Blue
Weight	Yellow

*Blank squares are tan

Name: Date:

Identify Shapes — Superhero

right	isosceles	equilateral	isosceles	isosceles	equilateral	isosceles	triangle	right	scalene
scalene	isosceles	hexagon	octagon	hexagon	hexagon	octagon	right	triangle	isosceles
triangle	right	rectangle	square	square	square	parallelogram	equilateral	triangle	right
trapezoid	equilateral	rectangle	rectangle	square	trapezoid	rectangle	right	right	equilateral
	right	isosceles	parallelogram	parallelogram	square	scalene	isosceles	right	scalene
									isosceles
triangle	octagon		octagon	hexagon	octagon		octagon		triangle
triangle	hexagon		hexagon	octagon	hexagon		hexagon	parallelogram	scalene
right	hexagon		pentagon	hexagon	pentagon		hexagon	octagon	hexagon
isosceles	hexagon						pentagon	octagon	hexagon

Key:

3 sided shapes	Blue
4 sided shapes	Skin
More than 4 sides	Black

*Blank squares are red

Name: Date:

Identify Shapes — Dinosaur

equilateral	equilateral	scalene	equilateral	isosceles	trapezoid		isosceles	triangle	isosceles
scalene	right	triangle	right	parallelogram	trapezoid	trapezoid	trapezoid	right	scalene
isosceles	equilateral	isosceles	triangle	trapezoid	trapezoid	parallelogram	trapezoid	isosceles	scalene
triangle	equilateral	isosceles	isosceles		rhombus	isosceles	isosceles	isosceles	scalene
right	scalene	equilateral	rectangle	trapezoid	rhombus	trapezoid	rectangle	isosceles	triangle
isosceles	isosceles			square	parallelogram	equilateral		isosceles	equilateral
right	rectangle	parallelogram	parallelogram	rhombus	trapezoid	isosceles	isosceles	isosceles	equilateral
rhombus	parallelogram	parallelogram	parallelogram	isosceles	isosceles	isosceles	triangle	triangle	equilateral
triangle	equilateral	scalene	rectangle		triangle	right	isosceles	equilateral	isosceles
pentagon	octagon	hexagon	hexagon	hexagon	octagon	hexagon	hexagon	pentagon	pentagon

Key:

3 sided shapes	Blue
4 sided shapes	Green
More than 4 sides	Brown

*Blank squares are black

Name: Date:

Identify Shapes — Sunflower

isosceles	scalene	isosceles	right	triangle	scalene	isosceles	equilateral	isosceles	triangle
equilateral	triangle	isosceles					scalene	scalene	equilateral
scalene	isosceles		rhombus	square	rhombus	parallelogram		scalene	triangle
scalene	isosceles		rectangle	rectangle	trapezoid	square		triangle	right
isosceles	isosceles		trapezoid	parallelogram	trapezoid	trapezoid		isosceles	isosceles
equilateral	isosceles	right					isosceles	equilateral	isosceles
equilateral	equilateral	right	isosceles	hexagon	octagon	equilateral	isosceles	triangle	scalene
isosceles	scalene	isosceles	triangle	right	hexagon	isosceles	triangle	isosceles	scalene
triangle	triangle	triangle	isosceles	scalene	hexagon	equilateral	right	right	right
hexagon	pentagon	hexagon	hexagon	pentagon	pentagon	hexagon	octagon	octagon	pentagon

Key:

3 sided shapes	Blue
4 sided shapes	Brown
More than 4 sides	Green

*Blank squares are yellow

Name: Date:

Identify Fractions — Heart and Arrow

4/6	4/6	4/6	10/15	6/9	6/9	2/3	2/3	2/3	4/6	4/6	6/9	6/9	4/6	10/15	8/12	8/12	2/3	6/9
10/15	8/12	2/3	2/3	6/9	2/3	10/15	10/15	8/12	10/15	2/3	10/15	8/12	8/12	4/6	4/6			10/15
6/9	4/6	4/6	8/12	10/15	8/12	4/6	4/6	8/12	8/12	8/12	4/6	8/12	10/15	2/3	4/6	9/9		10/15
6/9	4/6	4/6	8/12	8/12	5/15	4/12	2/6	2/3	6/9	10/15	3/9	5/15	5/15	8/12	12/12	8/12	8/12	10/15
4/6	4/6	6/9	6/9	2/6	4/12	4/12	5/15	1/3	8/12	3/9	3/9	4/12	1/3	9/9	4/6	6/9	6/9	6/9
2/3	2/3	2/3	5/15	1/3	2/6	1/3	2/6	4/12	1/3	4/12	5/15	5/15	9/9	2/6	2/6	8/12	10/15	4/6
6/9	4/6	4/12	4/12	2/6	5/15	4/12	5/15	1/3	2/6	2/6	2/6	2/6	2/6	5/15	3/9	4/12	10/15	2/3
10/15	2/3	1/3	4/12	3/9	5/15	3/9	2/6	5/15	2/6	1/3	3/9	2/6	1/3	5/15	1/3	5/15	2/3	8/12
2/3	6/9	1/3	1/3	2/6	2/6	2/6	3/9	1/3	2/6	2/6	4/12	4/12	2/6	2/6	2/6	2/6	4/6	6/9
8/12	8/12	1/3	5/15	1/3	2/6	2/6	4/12	3/9	2/6	3/9	4/12	1/3	4/12	2/6	4/12	2/6	6/9	2/3
4/6	6/9	10/15	2/6	2/6	3/9	1/3	3/9	2/6	1/3	1/3	1/3	3/9	2/6	3/9	5/15	4/6	4/6	6/9
8/12	8/12	6/9	1/3	4/12	4/12	5/15	9/9	2/6	1/3	4/12	1/3	1/3	1/3	1/3	5/15	8/12	8/12	2/3
8/12	2/3	4/6	6/9	2/6	2/6	12/12	5/15	4/12	4/12	1/3	2/6	3/9	4/12	1/3	4/6	10/15	6/9	8/12
10/15	10/15	6/9	4/6	6/9	9/9	4/12	2/6	1/3	2/6	3/9	3/9	5/15	5/15	4/6	4/6	6/9	8/12	6/9
8/12	2/3	8/12	10/15	3/3	6/9	3/9	4/12	5/15	2/6	3/9	2/6	3/9	6/9	2/3	8/12	4/6	2/3	4/6
6/9	6/9		15/15	6/9	2/3	4/6	1/3	2/6	3/9	4/12	2/6	8/12	2/3	10/15	8/12	2/3	10/15	10/15
8/12		15/15		4/6	6/9	4/6	8/12	1/3	5/15	5/15	4/6	6/9	2/3	6/9	2/3	4/6	4/6	2/3
2/3	6/9		6/9	8/12	2/3	4/6	10/15	10/15	3/9	4/6	10/15	8/12	10/15	8/12	8/12	8/12	2/3	8/12
4/6	4/6	4/6	8/12	8/12	8/12	2/3	4/6	8/12	2/3	4/6	4/6	2/3	2/3	4/6	10/15	4/6	4/6	10/15
6/9	10/15	2/3	6/9	8/12	8/12	10/15	6/9	6/9	2/3	4/6	8/12	8/12	2/3	6/9	10/15	4/6	8/12	2/3

Key:

Equal to 1/3	Red
Equal to 2/3	Blue
Equal to 1 whole	Gray

*Blank squares are white

Name: Date:

Identify Fractions — Spaceship

2/8	1/4	5/20	1/4	3/12	3/12	4/16	4/16	3/12	1/4	2/8	5/20	3/12	2/8	4/16	5/20	3/12	5/20	5/20
5/20	5/20	1/4	2/8	2/8	4/8	2/8	3/12	4/16	6/12	1/4	2/8	3/12	4/8	2/8	1/4	5/20	2/8	2/8
3/12	3/12	5/20	1/4	1/4	1/4	8/16	2/8	3/12	15/20	5/20	4/16	2/4	5/20	5/20	1/4	3/12	3/12	2/8
1/4	1/4	4/8	2/8	3/12	3/12	4/8	6/12	6/8	12/16	9/12	10/20	10/20	1/4	5/20	2/8	4/8	3/12	4/16
2/8	3/12	8/16	2/4	4/16	5/20	4/16	6/8				6/8	2/8	1/4	5/20	10/20	2/4	4/16	1/4
4/16	10/20	8/16	8/16	4/8	4/8	15/20	3/4	15/20	6/8	6/8	6/8	12/16	8/16	6/12	4/8	4/8	8/16	2/8
5/20	10/20	8/16	4/8	1/4	8/16	4/8	6/8	3/4	9/12	6/8	12/16	8/16	4/8	2/8	2/4	4/8	8/16	2/8
1/4				2/8	3/12	5/20	4/16	8/16	2/4	10/20	2/8	3/12	4/16	2/8				1/4
5/20	8/8	8/8	8/8	4/16	4/16	2/8	2/8	5/20	1/4	4/16	1/4	1/4	5/20	3/12	16/16	4/4	4/4	1/4
5/20	8/8	8/8	4/16	3/12	4/16	3/12	4/16	3/12	5/20	1/4	4/16	4/16	1/4	2/8	4/16	4/4	3/12	4/16
4/16	2/8	1/4	3/12	1/4	1/4	2/8	5/20	5/20	4/16	1/4	3/12	1/4	4/16	1/4	5/20	2/8	1/4	5/20
2/4	4/8	8/16	2/4	10/20	8/16	2/8	4/16	5/20	3/12	5/20	5/20	5/20	3/12	5/20	1/4	3/12	2/8	5/20
4/4	4/4	8/16	8/8	8/8	10/20	2/8	2/4	2/4	10/20	6/12	2/4	4/8	6/12	8/16	6/12	10/20	2/8	1/4
4/4	12/12	4/8	16/16	12/12	8/16	3/12	10/20	16/16	4/4	6/12	16/16	8/8	4/8	16/16	20/20	6/12	5/20	1/4
6/12	10/20	2/4	10/20	10/20	10/20	1/4	6/12	20/20	8/8	10/20	8/8	20/20	1/2	12/12	12/12	2/4	5/20	3/12
8/8	8/8	2/4	12/12	16/16	8/16	2/8	10/20	4/8	2/4	8/16	2/4	8/16	6/12	4/8	4/8	2/4	1/4	3/12
16/16	12/12	6/12	8/8	4/4	8/16	4/16	8/16	20/20	4/4	6/12	12/12	16/16	6/12	4/4	20/20	8/16	1/4	4/16
4/8	6/12	4/8	2/4	6/12	2/4	1/4	4/8	16/16	8/8	8/16	12/12	8/8	10/20	4/4	12/12	4/8	1/4	4/16
16/16	12/12	2/4	16/16	4/4	8/16	1/4	2/4	2/4	8/16	6/12	2/4	10/20	2/4	2/4	6/12	4/8	2/8	3/12
20/20	8/8	2/4	12/12	16/16	8/16	3/12	8/16	12/12	16/16	4/8	12/12	4/4	6/12	20/20	8/8	8/16	1/4	5/20

Key:

Equal to 1/4	Blue
Equal to 2/4	Black
Equal to 3/4	Gray
Equal to 1 whole	Yellow

*Blank squares are red

Name: Date:

Identify Fractions — Mouse

1/5	2/10	1/5	2/10	5/25	4/20	2/10	5/25	4/20	3/15	5/25	4/20	1/5	4/20	2/10	5/25	4/20	1/5	1/5
5/25	1/5	3/15	3/15	2/10	3/15	4/20	2/10	3/15	1/5	3/15	5/25	4/20	4/20	5/25	2/10	5/25	3/15	2/10
1/5	1/5	1/5	1/5	1/5	3/15	1/5	1/5	5/25	1/5	5/25	3/15	3/15	4/20	1/5	3/15	2/10	1/5	3/15
4/20	4/20	1/3	3/9	4/12	5/25	4/20	5/25	1/5	2/10	5/15	3/9	4/12	2/10	2/10	3/15	1/5	1/5	4/20
5/25	4/12	1/2	3/6	5/10	2/6	4/12	5/15	4/12	1/3	1/2	5/10	4/8	3/9	5/25	1/5	3/15	1/5	3/15
1/5	3/9	5/10	1/2	2/6	3/12	1/4	1/4	2/8	4/16	1/3	4/8	2/4	5/15	4/20	5/25	5/25	3/15	4/20
1/5	3/9	5/10	2/6	4/16			5/20			4/16	1/3	3/6	3/9	4/20	4/20	3/15	5/25	1/5
4/20	5/15	5/10	5/15			4/12	4/16	2/6			4/12	2/4	3/9	5/25	4/20	1/3	3/9	4/20
5/25	5/25	3/9	5/15			3/9	3/12	5/15			2/6	4/12	2/10	2/10	4/12	2/8	5/15	2/10
3/15	5/25	5/15	2/8			1/3	5/20	2/6			5/20	1/3	3/15	3/15	1/3	2/8	2/6	2/10
1/5	4/20	5/15	3/12			1/3	4/16	1/3			1/4	5/15	1/5	1/5	2/6	2/8	5/15	4/20
5/25	2/10	2/6	4/16	4/16			2/8			4/16	4/16	4/12	4/20	1/5	4/12	1/4	3/9	1/5
5/25	3/15	3/9	2/8	1/3	3/9	1/3	5/20	2/6	4/12	3/9	3/12	3/9	3/15	3/15	1/3	5/20	4/12	4/20
2/10	5/25	4/20	3/12	2/8	1/4	1/4	5/15	1/4	5/20	4/16	5/20	2/8	4/12	5/15	4/16	2/6	2/10	2/10
5/25	1/5	4/12	3/12	5/15	4/12	1/3	3/12	3/9	4/12	3/9	1/4	1/4	1/4	4/16	2/6	3/15	1/5	1/5
2/10	2/10	2/6	2/8	4/16	5/20	2/8	1/4	4/16	5/20	1/4	5/20	2/8	5/20	5/20	3/9	3/15	4/20	5/25
1/5	3/15	2/6	4/16	1/4	4/12	1/4	2/6	4/16	5/20	2/6	1/4	3/12	5/20	2/8	4/16	2/6	3/15	4/20
5/25	4/20	2/6	2/6	4/12	3/9	3/12	5/15	3/9	3/9	5/15	1/4	1/4	2/8	5/20	1/4	4/12	5/25	4/20
3/15	4/20	2/10	3/15	3/9	1/4	4/16	3/12	5/20	1/4	5/20	4/12	4/16	1/4	5/20	1/4	1/3	2/10	2/10
5/25	4/20	3/15	5/25	1/3	3/9	2/6	1/4	2/8	4/16	3/9	4/12	4/16	1/4	4/16	1/4	2/6	2/10	1/5

Key:

Equal to 1/2	Tan
Equal to 1/3	Black
Equal to 1/4	Brown
Equal to 1/5	Blue

*Blank squares are white

Name: Date:

Identify Fractions — Handheld

4/8	5/10	4/8	5/15	2/6	4/12	2/6	5/15	2/6	1/3	1/3	3/9	2/6	5/15	2/6	1/2	3/6	5/10	1/2
1/2	5/10	2/4	3/9	3/12	5/20	3/12	3/12	1/4	3/12	5/20	4/16	1/4	5/20	1/3	2/8	5/10	1/2	3/6
1/2	4/8	3/6	3/9	3/12	2/4	5/10	3/6	3/6	4/8	1/2	2/4	3/6	1/4	2/6	2/8	5/10	5/10	5/10
5/10	2/4	1/2	4/12	4/16	2/4	1/2	3/6		5/10	3/6	3/6	3/6	3/12	3/9	3/12	3/6	3/6	2/4
2/4	3/6	3/6	4/12	4/16	1/2	4/8				1/2	5/10	5/10	5/20	3/9	5/20	5/10	4/8	4/8
2/4	1/2	3/6	1/3	2/8	1/2	1/2	1/2	3/6	5/10	4/8	3/6	4/8	2/8	4/12	5/20	4/8	1/2	4/8
2/4	1/2	5/10	4/12	2/8	4/20	1/5	5/10	5/10	1/2	2/4	4/20	1/5	4/16	2/6	4/16	2/4	2/4	3/6
1/2	3/6	3/6	1/3	1/4	4/20	2/10	5/10	1/2	5/10	3/6	1/5	2/10	1/4	3/9	4/16	2/4	4/8	3/6
2/4	2/4	5/10	4/12	1/4	4/16	4/16	1/4	4/16	5/20	5/20	5/20	1/4	5/20	4/12	2/8	1/2	1/2	1/2
1/2	4/8	5/10	5/15	4/12	5/15	5/15	3/9	4/12	4/12	4/12	2/6	1/3	3/9	1/3	3/12	3/6	3/6	4/8
2/4	4/8	2/4	1/3	4/16	2/6	1/3	4/12	3/6	4/12	2/6	2/6	3/9	3/9	1/3	3/12	2/4	2/4	1/2
4/8	4/8	3/6	1/3	3/9	3/15	5/15	1/3	4/12	2/6	4/12	4/12	3/9		5/15	2/8	4/8	2/4	2/4
4/8	3/6	1/2	5/15	2/10	5/25	3/15	3/9	5/15	1/3	2/6		3/9	4/12	1/3	5/20	4/8	4/8	1/2
4/8	3/6	3/6	5/15	2/6	1/5	4/12	3/9	2/6	3/9	2/6	4/12	2/6	3/9	4/12	1/4	2/4	1/2	3/6
3/6	4/8	3/6	3/9	5/15	4/12	3/9	3/9	4/12	5/15	4/12	5/15	2/6	4/12	2/6	2/8	5/10	3/6	2/4
2/4	3/6	5/10	3/9	3/9	3/9	1/3	1/3	3/9	5/15	3/9	4/12	2/6	5/15	5/15	1/4	3/6	1/2	2/4
4/8	4/8	5/10	1/3	5/15	3/9	3/9	4/12	4/12	4/12	4/12	4/12	4/12	1/5	3/9	5/20	1/2	2/4	1/2
5/10	5/10	5/10	1/3	5/15	4/12	4/12	2/6	3/9	2/10	2/6	4/20	4/12	2/6	3/15	4/16	1/2	5/10	5/10
4/8	4/8	1/2	2/6	5/15	3/9	3/9	4/12	1/3	4/12	3/15	4/12	4/20	5/15	2/8	5/10	5/10	3/6	1/2
4/8	5/10	2/4	5/10	1/4	1/4	3/12	5/20	1/4	1/4	2/8	3/12	1/4	5/20	4/8	4/8	4/8	3/6	4/8

Key:

Equal to 1/2	Green
Equal to 1/3	Light Gray
Equal to 1/4	Dark Gray
Equal to 1/5	Black

*Blank squares are red

Coloring SQUARED

and

GRADE LEVEL MATH STANDARDS

Coloring Squared: Kindergarten can be used to supplement instruction toward fluency in:

Basic counting skills ones and tens

Understand the relationship between numbers and quantities

Understand whether a number is greater than or less than another number

Add and subtract within 5

Coloring Squared: First Grade can be used to supplement instruction toward fluency in:

Relate counting to addition and subtraction

Add and subtract within 20

Understand 2 digit numbers represent tens and ones

Add multiples of 10

Distinguish attributes of shapes

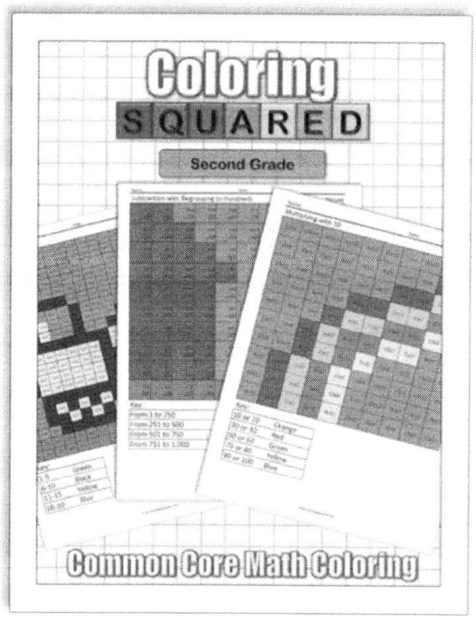

Coloring Squared: Second Grade can be used to supplement instruction toward fluency in:

Add and subtract within 20

Determine if a number is odd or even

Understand a 3 digit number represents hundreds, tens, and ones

Count within 1,000

Add and subtract with regrouping 2 and 3 digit numbers

Coloring Squared: Third Grade can be used to supplement instruction toward fluency in:

Round numbers to the nearest ten or hundred

Add and subtract with regrouping 2 and 3 digit numbers

Multiply and divide within 100

Understand shapes and their attributes

More great books from Coloring Squared!